Computer Interfacing

The Maplin series

This book is part of an exciting series developed by Butterworth-Heinemann and Maplin Electronics Plc. Books in the series are practical guides which offer electronic constructors and students clear introductions to key topics. Each book is written and compiled by a leading electronics author.

Other books published in the Maplin series include:

Audio IC projects	Maplin	0 7506 2121 4
Logic design	Mike Wharton	0 7506 2122 2
Music projects	R A Penfold	0 7506 2119 2
Starting Electronics	Keith Brindley	0 7506 2053 6

Computer Interfacing

Graham Dixey C.Eng., M.I.E.E.

BH NEWNES

Newnes
An imprint of Butterworth-Heinemann Ltd
Linacre House, Jordan Hill, Oxford OX2 8DP

A member of the Reed Elsevier group

OXFORD   LONDON   BOSTON
MUNICH   NEW DELHI   SINGAPORE   SYDNEY
TOKYO   TORONTO   WELLINGTON

British Library Cataloguing in Publication Data
A catalogue record for this book is available from the
British Library
ISBN 0 7506 2123 0

Library of Congress Cataloguing in Publication Data
A catalogue record for this book is available from the
Library of Congress

Edited by Co-publications, Loughborough
Typeset and produced by Sylvester North, Sunderland
— all part of The Sylvester Press

Printed in Great Britain by Clays Ltd, St Ives plc

Contents

Preface

This book is a collection of feature articles previously published in *Electronics — The Maplin Magazine*. In their original guise they comprised the series *Computers In the Real World*. They were chosen for publication in book form not only because they are were so popular with readers in their original magazine appearances but also because they are so relevant in the field of computing — a subject area in which it is evermore difficult to find information of a technical, knowledgeable, yet understandable nature. This book, we think, is exactly that.

This is just one of the Maplin series of books published by Newnes books, covering a wide range of Computing and Electronics topics. Others in the series are available from all good bookshops and from Maplin shops, and may be purchased from Maplin. In case of difficulty just dial 0933 410511 to check on price, availability and ordering details.

Maplin also supplies a wide range of computer hardware, computer accessories, electronics components and other products to private individuals and trade customers. Telephone: (0702) 552911 or write to Maplin Electronics, PO Box 3, Rayleigh, Essex SS6 8LR, for further details of product catalogue and locations of regional stores.

1 Introduction

The power of the microcomputer is evident to us all. Those who use computers a great deal, whether at work or at home, have accepted readily the interaction that takes place between human and machine. This interaction is performed through the medium of an interface, which is the term for any piece of hardware that makes possible the communication (in either direction) between the computer itself and its peripherals and, of course, a human operator. The word peripheral is also a general one used to describe anything that is connected to the computer but lies outside its periphery and so, although being essential for certain operations, is not usually an integral part of the computer itself. Figure 1.1 is a schematic diagram of a computer connected to a number of peripherals, in a context which many people might find familiar.

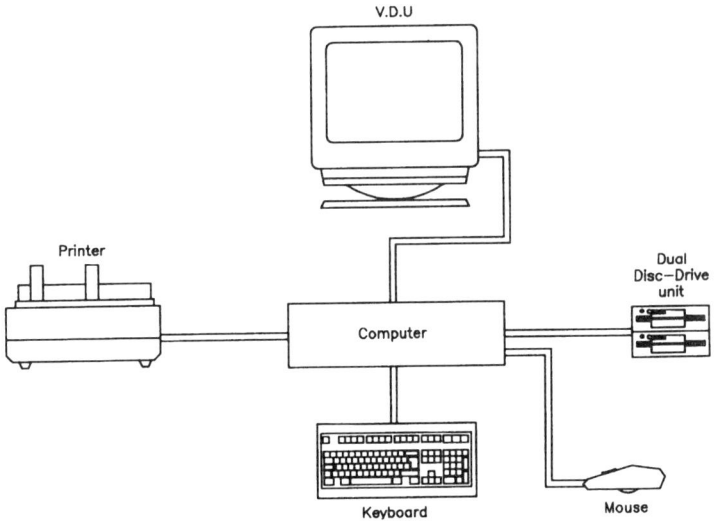

Figure 1.1 A computer and its peripherals. Not the only situation but one that is easily identified. The computer is seen as the central element, common to all computing situations; the number and nature of the peripherals can change according to the requirements of the system

The peripherals shown in this figure are: a visual display unit, or VDU (also known as a monitor); a keyboard; disk drive unit (hard or floppy); a printer and a mouse. Figure 1.1 merely shows, in a very general way, how the computer is central to its peripherals. It does not show the interfaces mentioned previously, nor does it indicate what the microcomputer actually comprises.

Processors

Figure 1.2 gives a rather more detailed insight into the component parts of a computer and the way in which the peripherals are interfaced. Here we see that the microcomputer consists of the processor itself (MPU for Micro Processor Unit), which is where all of the program instructions are decoded and acted upon, where the system timing originates that actually causes the program to run in the required sequence, plus other related tasks. Well known examples of processors are the 6502, Z80, 80386, 68040, Pentium and PowerPC.

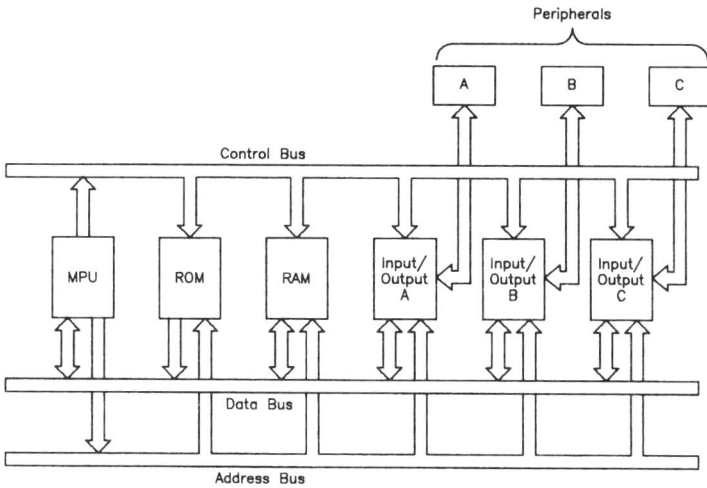

Figure 1.2   Essential block diagram for any computer, irrespective of its application. It's easy to see why the computer is often referrred to as a bus-oriented system. The input/output are quite general as shown and could refer to any of the peripherals and their interfaces shown in Fiqure 1.1

By itself the processor would achieve little if anything. To make it *do* something it must be given a program, which is nothing more than a list of instructions which, one by one, it obeys, like the slave that it is. This program may be stored in ROM (Read Only Memory) or RAM

Computer interfacing

(Random Access Memory, more usefully thought of as Read And write Memory). Most computers will have programs stored in both types of memory, those in ROM either being written to allow the computer to perform a variety of mundane functions such as reading the keyboard, writing to screen, or providing specific functions, such as running a high level language (BASIC, Pascal, and so on) or facilities such as word-processing. Programs in RAM are those entered by the user, usually from an external storage device such as a disk drive, although it is possible to enter short programs into RAM directly through the keyboard. The blocks marked ROM and RAM in Figure 1.2 indicate the presence of both types of memory, irrespective of how much of each is actually provided.

Interfaces and buses

The final blocks in Figure 1.2 are the interfaces to the peripherals themselves. Sometimes they are known as the input/output ports of the computer, especially when they are provided by a general purpose interface chip, such as the 6522 VIA (Versatile Interface Adaptor) or the Z80 PIO (Parallel Input/Output). Other interfaces may have names which, by familiarity, determine their likely function. For example, the Centronics parallel port is used with a number of popular printers, whereas some printers either require, or offer the option of, the use of a serial connection such as EIA 232 (also known as RS 232).

Connecting all of these blocks are the highways known as buses. They are the Address Bus that is used to identify and access the required area of memory at any instant, the Data Bus that is used to transfer data back and forth between processor, memory and peripherals, and the Control Bus whose functions include co-ordination of all the events during the running of a computer program.

It would be useful to define, in a general way, the function of any interface device. It is not unreasonable to ask why such a device is needed at all. To the initiated it may seem obvious; those with less experience of the ways in which computers and peripherals work may not find it so apparent.

Signal conversion

The need for an interface arises either because the signals originating in a peripheral are in some way different from those which the computer requires, or because the signals are the same but the speeds at which the computer and peripheral handle the data are quite different. Alternatively, some conversion may be needed just to allow effective communication between the two devices. This reasoning applies in either direction of data flow. It may be that the peripheral requires a signal that is quite different from the parallel digital output that the computer normally supplies. To take each of these points further, consider the differences that may exist between signals.

All signals may be classified as being either analogue or digital. An analogue signal is one which can take up any

Computer interfacing

of an infinite range of values and is capable of continuous variation. For example, in an audio amplifier the signal could have a value that was anything from a fraction of a millivolt to several volts; there is no one level that the signal can be assumed to have. The level at the pick-up end of the system is quite different from that at the speaker end. The same is true of any system that uses an analogue input and gives an analogue output. By contrast, a digital signal has only two values, known as logic 0 and logic 1. A system based on these values is termed a binary system; this is the system that all digital computers use. It is only necessary to assign nominal values of voltage to these two binary values and to design the computer circuits so that they are capable of distinguishing between them, a quite easy task for it to do. Naturally, if 0 V is assigned to logic 0 and +5 V is assigned to logic 1 (as is quite usual), we must expect that there will be departures from these values due, for example, to voltage drops in the system, but it is quite easy to allow the two logic levels to have tolerances and still be capable of being distinguished, one from the other.

Perhaps we can now begin to see how some of the problems arise. Suppose we have an analogue device whose output we wish to send to a digital computer for processing. The two types of signal are totally incompatible. In all probability the level of the analogue signal is far too small and furthermore the digital computer won't know what to do with it anyway. It becomes necessary to raise the level of the analogue signal (one of the processes known as conditioning) and to make an actual conversion from the analogue to the digital form. The circuit that performs this conversion is known as an analogue-to-digital converter (ADC). This situation arises in many industrial control systems where a computer is used to control a process, such as the temperature of a furnace, by monitoring the variable (in this case temperature) and generating a signal that is then sent to a controller to maintain the required temperature. The latter will often also need converting because it may well be an analogue device as well, quite unable to respond directly to the digital signal produced by the computer. The device that performs this conversion is known as a digital-to-analogue converter (DAC). This situation is shown in a simplified form in Figure 1.3.

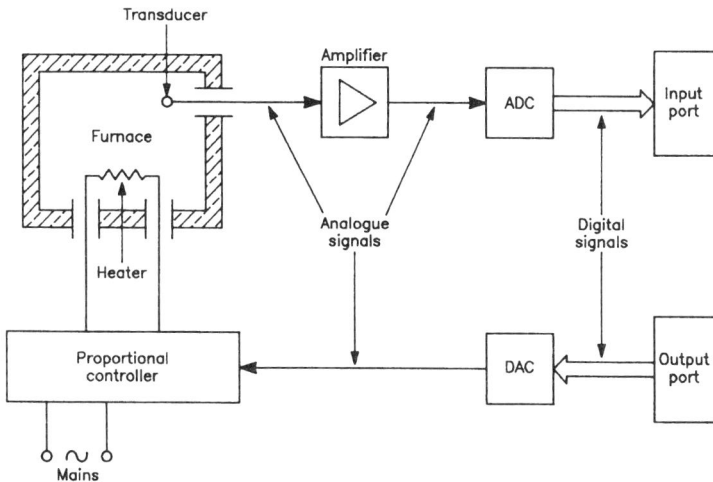

Figure 1.3   A practical application of a computer illustrating the need for signal conversion. Analogue output from transducer (monitoring furnace temperature) is amplified and converted to digital form before being processed by the computer. Conversely, digital output is converted to analogue form in order to control furnace power

Computer interfacing

A question of speed

The second case concerns two devices, the computer and a peripheral, that both handle digital signals but have quite different operating speeds. The best example is when a printer is driven from a computer. The latter is capable of working at quite incredible speed, often executing thousands of instructions in a few milliseconds. The printer, by contrast, is incredibly slow. Even the fastest printer is virtually at a standstill compared with the speed of the computer. Yet somehow it is essential to synchronise the two so that the printer gets its supply of data to carry out its task. An interface is needed that will keep the printer supplied with a stock of data (in what is called its buffer), which it can top up from the data held in the computer as required. The printer and computer will carry out what is called a handshaking procedure in order to do this. In a later chapter, the operation of various printers is described in detail.

The third case could also concern a printer but is equally applicable when connecting a computer into the public telephone network so that it can talk to other computers. The function of the interface now is to convert the normal parallel output of the computer into serial form. The difference between the two types of digital signal is illustrated in Figure 1.4. The parallel signal needs a separate line for each bit, whereas only a single line is needed, no matter how many bits the signal has, in the case of serial data. The special advantage of the latter is evident when long distance communication is the case, although it is obviously a lot slower than parallel transmission because the bits are sent one after the other.

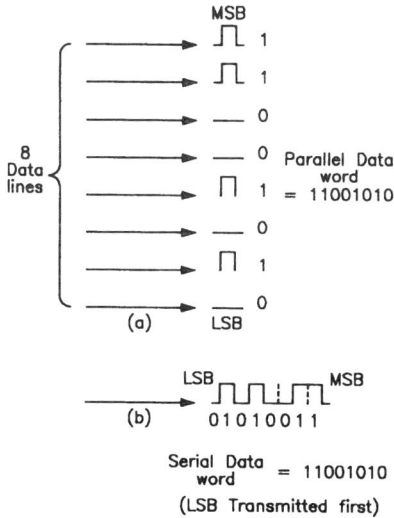

Figure 1.4 The difference between (a) parallel and (b) serial transmission of data. The parallel data needs one line per bit but all bits are sent at once. The serial data needs only one line, but the bits are sent one after the other

Also, extra bits have to be sent to indicate the beginning and end of a data word, as well as supplying information that allows errors to be detected and corrected. Some printers accept or even need serial input. Modems, used where communication over telephone lines is required, also have to condition the signal by using it to modulate the telephone carrier — the MO part of MOdem at the sending end — and to demodulate the telephone carrier at the receiving end — the DEM part of moDEM. Figure 1.5 shows two computers communicating through modems. Even here there can be incompatibilities. The rate at which the data is transmitted, known as the Baud rate,

Computer interfacing

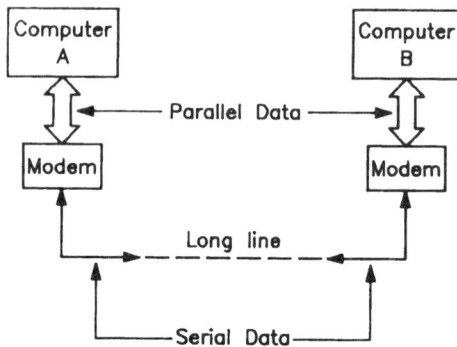

Figure 1.5 Two computers communicating via Modems. A conversion from parallel to serial form and vice versa is needed at both ends of the line

must be selected to be the same for both ends of the system, although it doesn't have to be the same in both directions. A more detailed discussion of the principles of modems follows in a later chapter.

Core task

In any computer system the computer is the central core. It should be evident by now that the computer consists solely of the MPU, memory, input/output chips and some additional logic. All of this can be accommodated on a single board, often of quite modest size. The VDU, keyboard, disk drive, and so on are not actually part of the computer proper but merely peripherals for it. It is quite possible to have a computer without a keyboard, without a VDU or any sophisticated display device and without any external storage device such as a disk drive unit. As long as it can run a program and perform some useful task it is still a computer. It will be designed to run a program in ROM as soon as power is applied to it and any variations on this program will be made by an operator setting a few external switches to selected positions. A rather hackneyed, but nonetheless readily appreciated, example is an automatic washing machine. The exact wash program run can be predetermined at the start by setting switches while the inbuilt ROM program looks after operations that, in general, are concerned with establishing time durations, water temperatures, switching motors on and off for washing and spinning, and so on. It doesn't look the least bit like the computers considered so far but it has exactly the same right to be called a computer. It is, for obvious reasons, usually referred to as being dedicated, since it never does anything else during its life than the one task. More conventional computers can be considered general-purpose machines.

All of this should have made it clear that, for one of several possible reasons, some form of signal conversion is often required. Hence the need for interfaces. What this book is about is (1) how we can make computers perform a variety of useful, and often quite different, functions for us, and (2) how the actual application of a computer is determined by the program run on it, the peripherals connected to it and the interfaces that provide the necessary degree of compatibility. Apart from the general interest that may be stimulated by discussions of the role of computers in the differing environments of the office, home, factory, hospital, and

Computer interfacing

so on, those readers who use a computer themselves may feel tempted to make wider use of it. To this end, where it is practicable or relevant, details are given for building interface circuits and experimenting with them and associated hardware. In this respect some hints on how to write the controlling software is also provided. As for who this book is for, it is for anyone interested in today's computer-orientated world. Only a modest amount of knowledge of electronics is assumed and, hopefully, not too much is taken for granted where computers themselves are concerned.

2 Transducers

If a computer is to interact with the real world, it has to be capable of exchanging data with it. This means that it will be accepting some form of data input, processing it in some way, then providing some form of output that is related to this input. Some of the possible peripheral devices that might be exchanging data with the computer were discussed in a general way in Chapter 1. In this second chapter we shall consider some of the devices that commonly provide data inputs. For the moment disk drives will be excluded as they will be dealt with in more detail later. Much of the present discussion will concern a device known by the general term of transducer. These are of particular importance in a way that the average computer user rarely has to contend with — since they are used to derive suitable signals from the natural environment. Hence they are largely concerned with the

Computer interfacing

application of computers in controlling a wide variety of physical processes rather than in the field of office automation, which is already well covered by the computing press.

However, there is one input device worth looking at briefly, at least from the hardware point of view, since it is the one that most people will have some practical acquaintance with. The device in question is the keyboard.

Most keyboards have two identifiable parts, the keyswitch array and the encoder. The keyswitches are usually simple single-pole single-throw types, with normally open contacts. The function of the encoder is to convert the simple switch closure into a complete ASCII code for that particular key. The output from the keyboard, generated in this way, is usually in parallel form, though a serial output can easily be obtained if desired. Other features of the encoder include suppression of switch contact bounce, ability to handle the case when more than one key is pressed at a time and generation of a pulse, usually known as the keyboard strobe, which is output whenever a key is pressed and signals this salient fact to other circuits. A further feature of the encoder is that it allows a key to have several different functions by combining it with the use of shift or control keys.

The ASCII code (American Standard Code for Information Interchange) mentioned here is actually a 7-bit code but the eighth bit of the byte is often used for parity, which is a method of error checking. Thus, if a particular 7-bit code has an odd number of 1s, making the parity

14

bit a one then produces even parity (an even number of 1s in the byte); conversely, if the parity bit has a value that produces an odd number of ones in the byte, then the result is odd parity.

The most popular key layout is the standard QWERTY arrangement, so called because these are the first six in the upper left area of the alphabetic keys. To make keyboards user-friendly keys should be of a certain minimum size, be arranged on standard centres and have some sort of tactile feedback for the user, usually a result of positive travel and a sensation experienced through the finger tips that data really has been entered. Zero travel, membrane-type keyboards have, in the past, been something of a failure. Staggered, sloped or dished arrangements are common on computer keyboards. These factors are mentioned because they influence enormously the ease with which a human can interact with a computer, especially over prolonged time periods.

Rollover and lockout are ways of preventing incorrect codes from being generated when more than one key is pressed at a time. In *n-key lockout* the first key down generates the strobe, the others do nothing. Although very simple it has the obvious disadvantage that key presses can be missed completely. This snag is avoided with *2-key rollover* in which, if a second key is down at the same time, logic is used to allow the keypress of the second key to be recognised soon after that of the first key down. There is also *n-key rollover* in which the codes for any number of keys down are successively stored and recognised in turn — rather too complex for most applications.

Computer interfacing

Encoder circuits

As already stated the function of the encoder is to convert the single pole key press into the equivalent ASCII code for that key. There are various types of encoder but one of particular interest is the scanning encoder. In this type the keys are arranged in a matrix. For example, a 64-key keyboard could be arranged on an 8 x 8 matrix. This is not the physical arrangement of keys, of course; they are laid out on the QWERTY system as already discussed. The matrix refers to the way in which the below board wiring is connected to the switches. The wiring is laid out as 8 columns x 8 rows with one switch across each column/row intersection (see Figure 2.1). Thus, pressing any key produces a unique short-circuit at a given intersection. A key press is identified by a scanning process that interrogates each key in turn through an oscillator and a decoder/selector circuit. The elements of a keyboard based on these principles is shown in block diagram form in Figure 2.2.

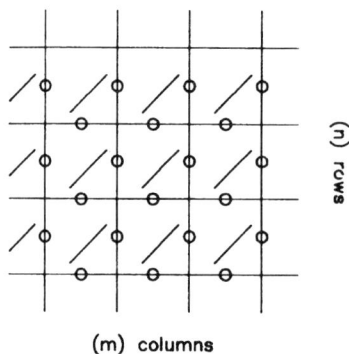

Figure 2.1 Part of an m x n keyboard matrix, showing positions of keyswitches on the row/column intersections

In operation a gated oscillator, running at a frequency of about 50 kHz, clocks a 6-bit counter. The top three bits of the counter are decoded to a 1-of-8 sequence that places a logic 1 on each of the eight column lines in turn. The bottom three bits of the counter drive a 1-of-8 data selector that interrogates each of the eight rows in sequence. When a key is pressed the coincidence of the active column and the row where the key press has occurred is used to stop the oscillator while the code for the key is generated; a keyboard strobe is also produced. For the circuit shown in Figure 2.2 the actual code generated is simply the pattern of binary digits that appears on the seven output lines at the instant the counter is on hold. This figure also includes extra gating to make use of the shift and control keys to generate extra codes.

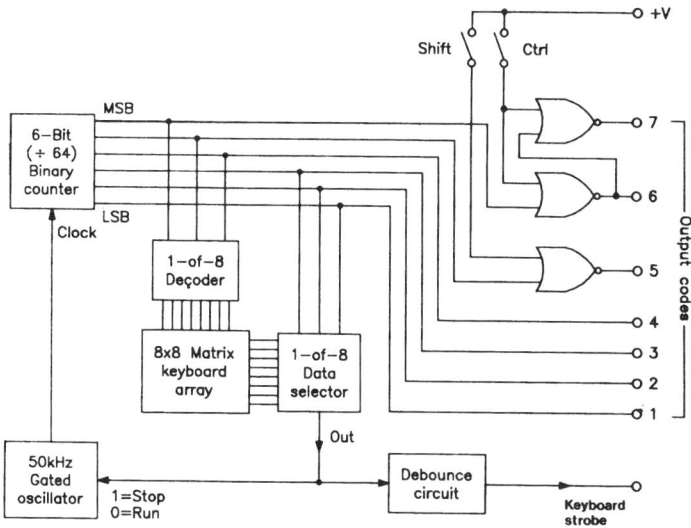

Figure 2.2   A scanned keyboard

Computer interfacing

Any desired code can be generated, for example the ASCII code referred to, by storing the codes in a ROM and using the binary outputs from the scanned keyboard, not as the actual codes themselves, but as addresses in the ROM where the actual codes are found. The block diagram of Figure 2.3 shows the arrangement for a scanning keyboard which uses codes stored in a ROM, each output code from the ROM being held in a set of latches before being passed to the computer.

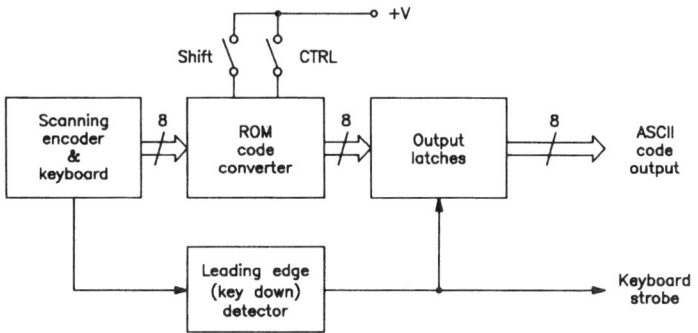

Figure 2.3   A scanned keyboard with ROM to store the ASCII codes

Transducers

A transducer, in the most general sense, is merely a device able to convert one form of energy into another. There are a number of everyday examples of this conversion process. The loudspeaker is one, in which an electrical input is converted into an acoustic output. The microphone performs the opposite process, producing an electrical signal from an acoustic input. In general, as a computer requires an electrical input of a very specific kind, the types of transducers to interest us will be those that take some physical quantity and derive an equivalent electrical signal from it.

It is worth considering some examples of physical quantities we might wish to convert in this way, and consider how such conversion might be carried out. Their relevance to computer control is also worth investigation.

Examples of such physical quantities are: heat, light, weight, force, linear displacement, angular displacement, stress in mechanical structures, and fluid velocity.

Transducers for all of these quantities exist and may be classified as being either active or passive types. To make the distinction clear, an active transducer generates electrical energy directly — an electromagnetic microphone is an example, as it uses the principle of electromagnetic induction. By contrast, the passive transducer requires an electrical supply which it then modifies in some way so as to develop its output. A capacitor microphone is an example of a passive transducer which needs to be fed with a d.c. supply; the moving plate of the capacitor then modulates this d.c. supply with the acoustic signal.

Transducers for temperature measurement and control

An industrial oven, furnace or similar environment can be readily controlled by a microcomputer. Transducers exist that allow a voltage proportional to temperature to be obtained. Two possible types are the thermocouple and the resistance thermometer.

Computer interfacing

Thermocouples are temperature-sensitive devices that operate on the Seebeck effect. When a circuit is formed of two dissimilar metals (Figure 2.4) and heat is applied to one of the two junctions, a voltage or current is developed in the circuit proportional to the difference in temperature between the two junctions. Examples of metal combinations that are commonly used are: copper/constantan and iron/constantan. The useful temperature range for these is −250°C to +400°C and −200°C to +850°C respectively. As these devices generate electricity directly they are, of course, active transducers.

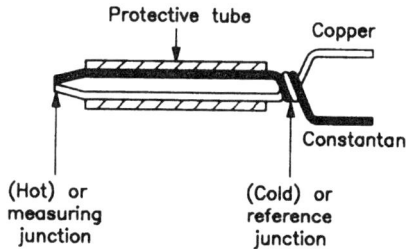

Figure 2.4   Essential construction of a copper/constantan thermocouple

The resistance thermometer operates on the principle that metals change their electrical resistance with changes in temperature. Resistance thermometers are particularly useful where high sensitivity is required. They are used to measure temperatures from approximately −75°C to +550°C. Platinum or nickel are the materials commonly used.

An inexpensive temperature transducer of the resistance thermometer type is the thermistor. It is made from

20

sintered mixtures of metal oxides and has semiconductor characteristics. It has a negative temperature coefficient, which means that an increase in temperature causes a reduction in the resistance of the device. The operating temperature is fairly limited ($-100°C$ to $+400°C$). Typical values are a resistance of 5000 ohms at $0°C$ and 100 ohms at $150°C$. One problem with thermistors is the extreme non-linearity of their characteristics but they are very sensitive indeed and, in the bead variety, very fast acting.

Both types of resistance thermometer are passive devices and are generally used in a Wheatstone bridge circuit in order to develop a voltage related to the resistance change — the bridge out-of-balance voltage. Since the actual transducer may be at a distance from the bridge itself in a practical situation, the possibly long connecting wires would become a fixed part of the transducer total resistance; to avoid this drawback, a modified form of bridge, called the Mueller bridge, is used instead. This is shown in Figure 2.5.

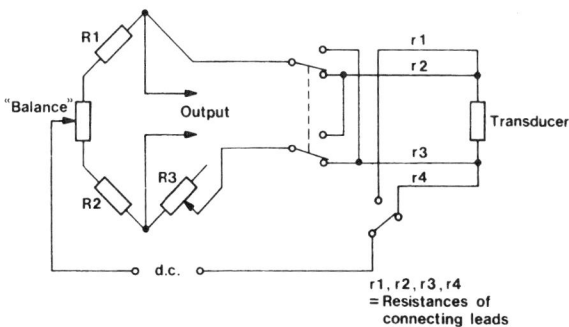

Figure 2.5   The Mueller bridge, used when long connecting leads are involved

Computer interfacing

Transducers for force and displacement

The strain gauge

The strain gauge works on the principle that resistance of a conductor depends upon its physical dimensions. Thus, if something influences any of these dimensions, a change in resistance will occur. This change can be detected and measured in a bridge circuit. These devices can be made from various metals, alloys or semiconductors. In practice they take the form of a very fine conductor network that may be unbonded (that is free in space) or bonded to a fine foil which is cemented to the member in which the force is to be measured.

Typical resistance values are 120, 350 and 1000 ohms. It is quite usual to place two similar strain gauges in opposite arms of the bridge, one being merely a dummy to give a reference to which the other, under stress, is referred.

The linear variable differential transformer (LVDT)

This is a transducer that can be used to measure linear displacement directly and force with a modification. Its construction is shown in Figure 2.6. A single primary winding is surrounded by two identical secondary windings. A magnetic core moves down the centre of the coil arrangement and will influence the amount of electromagnetic coupling between the primary winding and each secondary winding, according to its position relative to them. If exactly centrally placed the coupling is the same to both secondaries, and the induced voltages

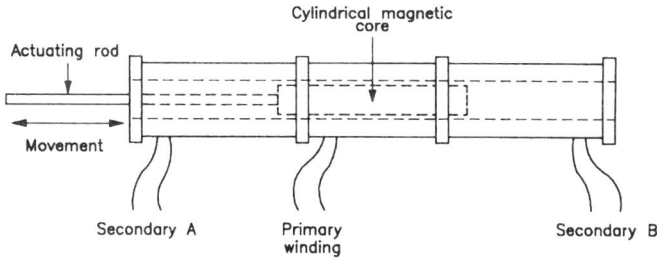

Figure 2.6 Basic construction of a Linear Variable Differential Transformer (LVDT)

are equal and opposite. Net output is thus zero. Movement in either direction causes one or other secondary voltage to be larger than the other; so there is a net output.

Not only is magnitude of the displacement sensed but the direction also, as there is a phase reversal as the null point is passed through. The a.c. output can be passed to a phase sensitive detector which rectifies it and produces a d.c. output that is either positive or negative (or zero) depending upon the direction of movement of the core (see Figure 2.7).

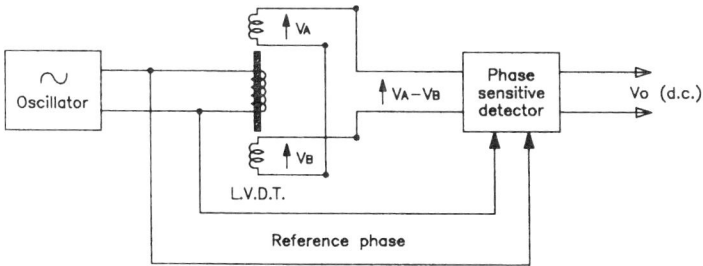

Figure 2.7 A polarity-sensitive output, proportional to linear displacement, can be obtained from a LVDT using a phase sensitive detector

Computer interfacing

Optical transducers

The photo-conductive cell

One of the most useful of optical transducers is the cadmium sulphide (CdS) photo-conductive cell. This has a clear plastic case containing a resistance element whose value depends upon strength of light falling on it. Enormous variations in resistance are possible, from tens of kilohms down to a few hundred ohms being typical. The coefficient of resistance change with light level is a negative one but is sufficiently constant to give reasonable linearity between the two quantities.

Optically coupled isolators

An optical coupler (Figure 2.8) consists of a gallium arsenide (GAs) infra-red emitting diode and a silicon photo transistor mounted in close proximity but electrically isolated. Such a device has the advantages of very high electrical isolation between input and output, good

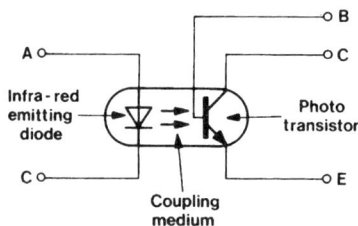

Figure 2.8 The optically-coupled isolator (opto-coupler or opto-isolator)

24

linearity between input and output currents, compatibility with TTL logic circuits, high speed, long life, good mechanical strength and a high current transfer ratio. It can be used as a simple switch or in a linear mode.

In switching mode the photo transistor operates under saturated conditions and switches between its on and off states. In certain industrial applications of digital equipment it is necessary to interface between sensors, such as microswitches situated in electrically noisy environments, and the control equipment itself. The circuit of Figure 2.9 shows how this can be done with an optical coupler, so avoiding the pickup of extraneous noise.

The optical coupler can also be used to transmit digital data between two digital systems where there is a substantial difference in the earth line potentials. This is shown in Figure 2.10. The data rate is normally limited to about 125–150 kHz.

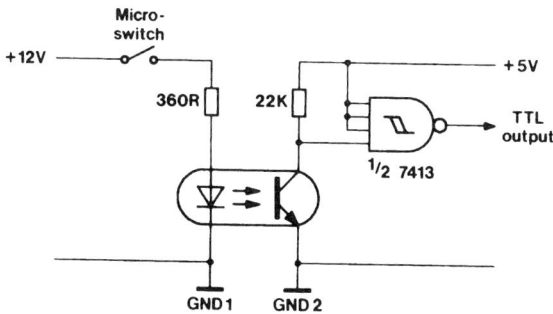

Figure 2.9 Using an opto-isolator to interface a microswitch to a digital circuit, avoiding ground noise problems

25

Computer interfacing

Optical couplers also exist as multi packages, e.g. quad isolators, with high gain, high speed circuitry and with triacs as the output switched element.

In linear mode the linear relationship between diode forward current and photo transistor current is used. In this mode the input current may be a function of some other variable, e.g. resistance change, and the isolator is used purely to obtain linear coupling with a high degree of electrical isolation.

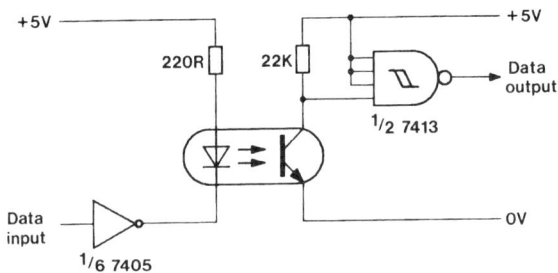

Figure 2.10   Using an opto-isolator to couple digital data between two circuits

Two other ways of using optical transducers are shown in Figure 2.11. In Figure 2.11(a) an optical coupler is arranged so that a slotted disc passes between the transmitting and receiving sections. Each time the slot appears in the optical path an output pulse is generated.

26

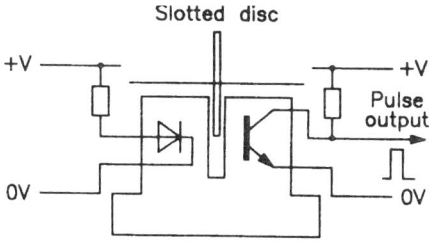

(a)

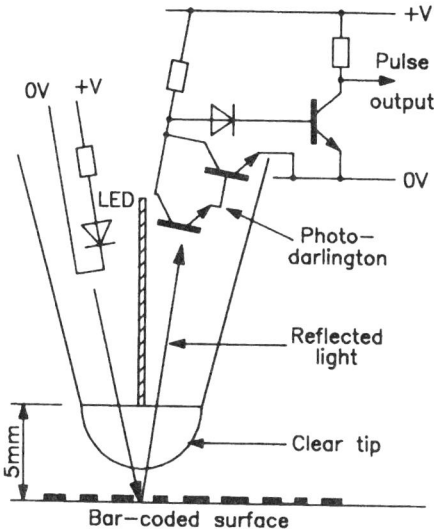

(b)

Figure 2.11   (a) Principle of optical coupling applied to the generation of pulses by a rotating, slotted disc (e.g. to measure engine speed), (b) Using a LED and a photo-Darlington amplifier to read bar codes

Figure 2.11(b) shows a reflective transducer which generates a serial digital output as it moves over a succession of black and white bars (e.g. bar codes). The distance of 5 mm is optimum and can be kept constant by a clear plastic lens at the tip, which contacts the surface. For high gain a photo-Darlington pair is used to drive a BC108 output stage.

The photo-diode as an optical sensor

The advantage of a photodiode over a CdS cell is its very much faster response, measured in microseconds or even nanoseconds. The silicon photo-diode is normally operated with a reverse bias so that, in the dark, the diode current is extremely small, a few tens of nano-amps only. An increasing light intensity causes a nearly linear increase in diode current. The circuit shown in Figure 2.12 gives an output voltage of 1 V/10 nA of diode current and allows the output to be read by a microcomputer by strobing pin 8 via the 1N914 diode. The low diode currents involved necessitate the use of a CMOS op-amp and correspondingly high values of the input and feedback resistors.

Potentiometric transducers

Potentiometers, whether linear or rotary, are able to give an output voltage which is a function of the wiper position. Thus, they are useful in providing a position dependent voltage. Various applications may spring to mind, including position feedback in control systems, liquid level measurement (by using a float and a suitable mechanical linkage to drive the wiper) and input demands from front panel controls. The accuracy of the derived signal depends upon the linearity of the potentiometer track (resistance variation with change of wiper position) and resolution (the number of turns of wire used to make up the track — the finer the wire, the more fragile but the better the resolution).

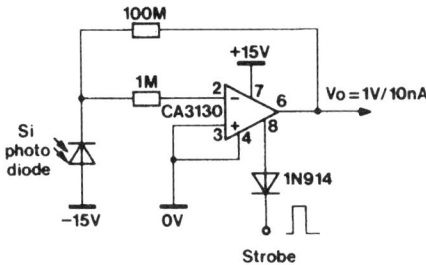

Figure 2.12   The silicon photo-diode used to measure light level. Good linearity exists between output voltage and ambient light level. The response is fast too

There are a variety of other transducers, for example for measuring flow rate of liquids, for measuring pressure, for measuring the pH factor of various liquids, humidity, density, and so on. They all have one thing in common;

they all take natural quantities as their input and produce an analogue output. However, there is one type of transducer that is capable of producing a digital output directly — the digital shaft encoder.

The digital shaft encoder

This is the only true transducer capable of giving a parallel digital output. It comprises a circular disc on which a series of radial patterns of light and dark areas, each representing a binary number, is applied, to be read by a number of sensing heads. Each of these binary numbers corresponds to a unique angular position of the disc. Thus, since an 8-bit number can take any of 256 different values, an 8-bit disc (that is, having eight annular tracks) can identify 256 separate angles, each being 360/256 = 1.4° apart. In practice, resolution usually needs to be better than this. The sensing heads may be of the contact, magnetic or optical type.

In the contact type, brushes bear on the annular tracks, either making electrical contact or not according to whether a binary 1 or binary 0 is being read. This type is subject to the usual problems of such contact arrangements, namely physical wear, dirt, friction, arcing, and so on.

The magnetic type offers an improvement by using a magnetically coated disc on which the binary patterns have been pre-recorded. Readout is effected by small toroidally wound heads.

Best of all is the optical type of sensing head. This gives the best accuracy and imposes no mechanical loading on the disc. The disc is usually photo-etched so that it has clear and opaque regions to represent binary values. The light from a suitable source then either passes through the disc or not, this being detected by means of photo-cells. A typical trackwidth is about 12 microns (a micron is a millionth part of a metre) and 14-bit optical encoders are common. With this number of bits the resolution is 214 = 16,384 angular intervals, each of value 360/ 16,384 = 0.022°.

The binary pattern used on disc encoders does not follow a pure binary sequence, as if it did large angular errors would occur whenever the sensors stopped on the junction of two segments. This is because the sense heads always tend to read the total number of logical 1 bits and form the output from this. Thus, a sense head on the junction of 0111 (7) and 1000 (8) would actually read 1111 (15), giving a substantial error. However, if a code is used in which successive binary numbers differ by only a single bit, the error is very small. Such a code is called a Gray code. In this code decimal 7 is 0100 and decimal 8 is 1100. Sensors reading the junction of these two numbers now read 1100, thus simply rounding up to the larger value of the two.

This survey of transducers has necessarily been brief but perhaps has given some idea of the wide variety that are obviously available. The next stage is to consider what to do with their outputs before their data can be processed by the computer. In almost all cases the output is in analogue form, and usually it is of quite a small magnitude. What has to be considered next is how to condition this analogue signal and how to convert it to digital form so that it is acceptable to a microcomputer.

3 Conversion between digital and analogue

We saw in the last chapter that signals generated in the real world to represent the quantities that we wish to measure and control are usually quite different from those that the computer requires. In short, they are analogue signals whereas the computer requires a digital input. In order to make the computer accept data that represents the input quantity, the analogue signal has to be converted into the appropriate digital form, for example, 8-bit binary. Similarly the only type of signal that the computer is able to *generate* directly for control purposes is a digital one. If the control device is an analogue type, as it often is, then the computer's digital output must be converted to an analogue form accept-

Computer interfacing

able to the analogue controller. There are exceptions, such as stepper motors which can generate linear or rotary movement in direct response to a digital signal from the computer. The principles and applications of these are discussed later.

The two types of conversion process mentioned above are known, respectively, as: analogue-to-digital conversion, requiring the use of an analogue-to-digital converter (ADC): and digital-to-analogue conversion, requiring the use of a digital-to-analogue converter (DAC). There's a number of different principles involved and, hence, a variety of possible circuits for both types of converter. Some of these are now discussed, as well as how they themselves interact with the computer in order to pass the required data back and forth.

Sampling the input

There is a tendency in a discussion of this sort to think of the analogue quantity as being constant in value, an assumption that may not be true. In fact, even when the quantity being measured and controlled is temperature, this is bound to vary somewhat even if at a very slow rate. Some quantities, such as velocity and acceleration, may vary extremely rapidly. If, at some instant in time, the analogue input is converted to a digital equivalent, all that has actually been done is to express the signal's digital value at that instant only; it may well be quite different a short time interval later. Figure 3.1 should make this quite clear and also illustrate why the analogue input must be converted at successive instants of time so as to keep track of the varying nature of the signal.

Conversion between digital and analogue

Note that several conversions are made for every cycle of the input waveform; a sinewave input is shown in Figure 3.1, but the principle is correct whatever the nature of the varying analogue input voltage. This process of converting samples of the input at successive times is, naturally, called sampling. The more samples taken during the period of the input, the more accurately does the digital data represent the nature of the analogue signal.

Obviously there is a limit to the number of times a signal can be sampled in any given time period; also, the higher the frequency of the analogue input, the more difficult it becomes to sample it often enough.

Quantisation

Because a digital signal changes in a *series* of steps rather than continuously as does the signal analogue, the digital values will quite often not coincide exactly with their analogue counterparts. They will approximate to them

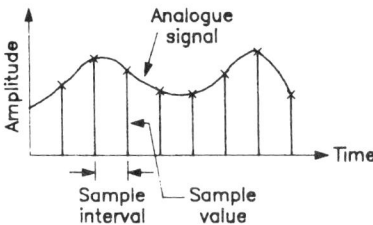

Figure 3.1   An analogue signal is sampled at regular instants of time, these samples being individually digitised

Computer interfacing

in some degree, the closeness of the approximation being determined by the number of bits of the digital value.
For example, if the converter has only four bits, there
are only 16 possible digital values. This situation is
shown in Figure 3.2. Each of the dotted lines corresponds
to a digital value being exactly equivalent to the analogue
value plotted horizontally. The process of approximat-
ing to an analogue voltage in this way is called
quantisation and the discrete levels are called
quantisation levels.

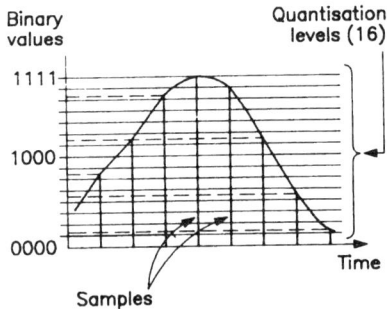

Figure 3.2   The heights of the samples may not always
correspond exactly to a binary value; approximation then occurs

The larger the number of bits, the more quantisation levels there will be and the better the approximation. For
example, an 8-bit converter will have 256 quantisation
levels, a 10-bit converter will have 1024 levels and so on.

Conversion between digital and analogue

Full scale range, quantisation levels and resolution

A moment's thought should reveal that, as the analogue signal always has a maximum value, this in turn must correspond to a maximum digital value. For example, an 8-bit converter may be designed so that the maximum value of the analogue input is limited to +5 V. Then the lowest digital value (00000000) will logically correspond to an analogue voltage of 0 V. At the other end of the scale, the full digital value of 11111111 will correspond to +5 V. From this one can easily work out the analogue voltage difference between one digital value and the next, known as the quantisation interval.

For example, if there are 256 quantisation levels (including zero) there are 255 steps between them. If the full scale range of the analogue voltage is from 0 V to +5 V, the quantisation interval will be 5/255 = 0.0196 V or 19.6 mV. The practical significance of this is that it expresses how closely one can get to an accurate equivalent of the analogue input. The term used to describe this is resolution. Resolution can be expressed as a fraction e.g. 1 in 255 or as a voltage as above.

It is interesting to consider what happens if an analogue signal is first put through a process that converts it into a series of digital samples and then put through the reverse process that converts these samples back into an analogue signal. Would you expect to end up with an exact replica of the original analogue signal? Figure 3.3 shows that this is certainly not true, especially if the converter uses only a small number of bits. In this figure a

Computer interfacing

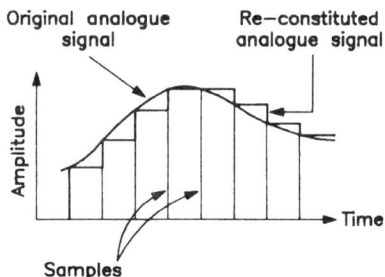

Original analogue signal / Re-constituted analogue signal

Figure 3.3 Showing the type of error that occurs when a digitised analogue signal is reconstituted

series of digital samples of a sinewave are taken which, when converted back, produce a corresponding series of steps; this is the re-constituted analogue waveform. A quite marked difference is evident between the original and re-constituted waveforms. There is an amplitude fluctuation now that was not present in the original signal; this higher frequency variation is given the name, *quantisation noise*.

Sample-and-hold circuits

The reason for needing sample-and-hold circuits can be appreciated by asking the question, how long is an instant of time? It is rather like the old question, how long is a piece of string? Obviously there is no specific answer. The relevance of this question is that any

38

conversion process takes a finite length of time. While this conversion is taking place the signal may well be varying in amplitude. It is hardly reasonable to ask the converter to digitise a sample at a specific instant of time then to change that value during conversion! The answer is to freeze the sample prior to conversion, this process taking negligible time compared with the conversion time that follows. Then the converter does its job and produces the digitised sample, after which the analogue sample is released and a new sample frozen. The reason for calling the circuit that performs this vital task a sample-and-hold circuit should now be obvious.

Figure 3.4 shows a sample-and-hold circuit in which an FET switch is used to connect the buffered analogue input voltage to the hold capacitor during the brief duration of the sampling pulse. The hold capacitor is also buffered in order to reduce the possibility of leakage during the conversion period.

Signal conditioning

This is the term given to a process carried out on the analogue signal prior to converting it into digital form. Under this heading may be included such processes as linearising, offsetting, noise-reduction and amplification. Lack of space precludes a discussion of all these but Figure 3.5 shows a circuit for an instrumentation amplifier which raises the signal level to that required in order to achieve the full-scale value referred to previously. As it is a differential amplifier it will, in fact, provide the last

Computer interfacing

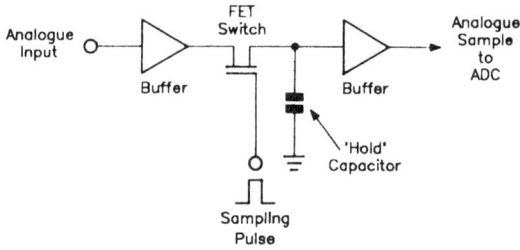

Figure 3.4   A sample-and-hold circuit

three types of conditioning listed. The full-scale analogue voltage is related to the maximum analogue signal to be handled by the formula given in this figure. With this circuit it is feasible to obtain any voltage gain value between unity and 10,000. The resistor pairs R2 and R5, R3 and R6 and R4 and R7 should be matched as accurately as possible. This leaves resistor R1 as the one most convenient for varying to obtain the required gain.

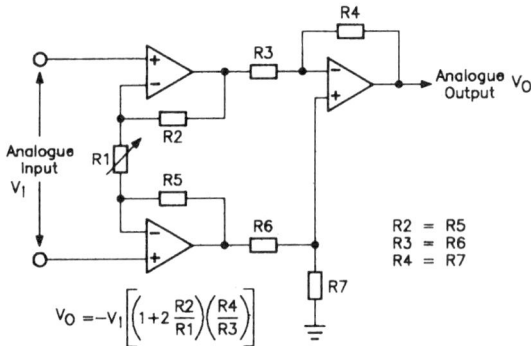

$$V_O = -V_I \left[\left(1 + 2\frac{R2}{R1} \right) \left(\frac{R4}{R3} \right) \right]$$

R2 = R5
R3 = R6
R4 = R7

Figure 3.5   Typical instrumentation amplifier used to condition an analogue signal prior to sampling

Conversion between digital and analogue

Multiplexed inputs

In instrumentation applications of computers there may be a number of transducers whose outputs have to be monitored by the computer with either analysis or control to follow. A deep space probe, for example, would have many parameters to monitor and it is hardly feasible to have a separate computer for each; nor is it even necessary or desirable to provide separate ADCs for each input. What is necessary is to make each transducer link up to the computer in turn. Thus, sampling is seen at two levels. Not only is each analogue signal sampled continuously with respect to time, but each transducer is also sampled in turn. This process is known as multiplexing. Thus, if four transducers are known as A, B, C and D, and their samples numbered as they appear, they could be given the identities A1, B2, C3, D4, A5, B6, C7, D8, A9, and so on. Figure 3.6 shows the idea in schematic form. The multiplexer shown in this figure contacts four switches (electronic of course) which are closed, one only at a time, in turn be a decoder circuit which is driven from a counter, the latter providing the channel addresses in sequence. There isn't anything very difficult about this idea of addresses. Most likely they would just consist of the binary sequence 00, 01, 10, 11, these being easy to generate and decode.

The converter-computer linkup

The object of an ADC is to provide the computer with digital data at its input/output ports. Such conversion may, however, be quite a slow process, certainly when

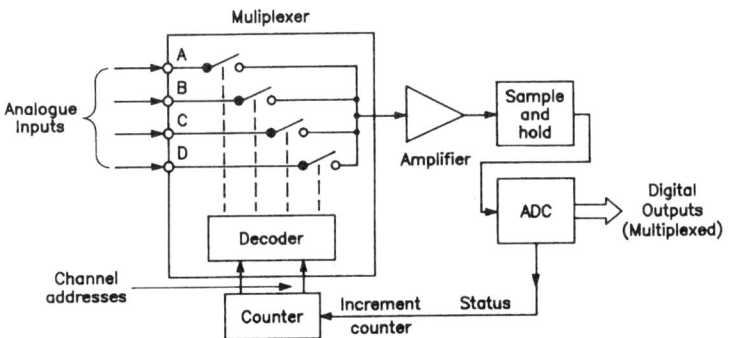

Figure 3.6 Multiplexing four analogue inputs is economical in terms of the hardware requirements

compared with the rate at which the computer can handle data. This makes it necessary to synchronise the two rates in some way. This may be done by linking the ADC to two available input/output lines of the computer, these two lines having control functions known as start convert and status. These lines allow the converter and computer to perform a handshaking procedure as follows.

The computer initiates the conversion process by sending a pulse to the ADC, on the start convert line; this is readily generated by software. At this time, when the conversion starts, the status line is taken to a predetermined logic level (either 0 or 1 depending on design) to signify that the ADC is now busy. At the end of the conversion process, the logic level of the status line inverts to indicate it has now returned to the ready state. The question is, what is the computer doing while the conversion is being carried out? There are two possibilities;

either the computer is doing something else — in effect multi-tasking — or it is doing nothing, such as waiting in a loop for the ready signal to appear on the status line. In the former case a special arrangement has to be made for the computer to leave its other work at the end of conversion in order to accept the converted data. This can be done by using an interrupt procedure something that will be considered in detail in the next chapter of this book.

The above discussion sets the background for a look at some of the variety of circuits used to convert, either way, between the analogue and digital forms of signals.

Digital-to-analogue converters

The process of converting a digital signal to an analogue one is relatively simple, both in principle and hardware design. There are two main methods, both being shown in Figure 3.7.

The method of Figure 3.7(a) relies upon using resistors whose values are weighted according to the columns of the binary value to be converted. Actually, the resistor values are in inverse ratio to the weightings of the binary columns, so as to produce currents flowing into the junction that are in direct proportion to the column weightings. Therefore, as a 4-bit binary number is weighted 8421, the current I_1 in resistor R1 is twice the current in R2, four times that in R3 and eight times that in R4, for the cases when the input is a logic 1. The output voltage V_0 is proportional to the binary number

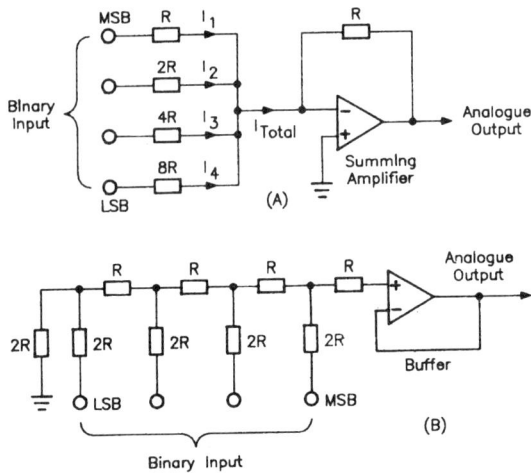

Figure 3.7 (a)The weighted resistor type of DAC and (b) the R, 2R DAC, both shown as 4-bit converters

input, thus achieving the required conversion. The op-amp is used as a summing amplifier in this example. Although it may not be an obvious drawback, this circuit has the disadvantage of requiring a wide range of different resistor values, especially with 8-bit and 10-bit converters. This would not matter so much if the design was intended to be built discretely, but present practice is to use thin-film or thick-film resistor techniques.

The R,2R ladder network DAC of Figure 3.7(b) overcomes this drawback. As its name implies it uses only two different resistor values, with a simple 2:1 relation between them, no matter how many bits are converted. Only by an involved network analysis can it be shown that this circuit actually works (or by practical experiment of course!). The role of the op-amp in this case is to act as a unity gain buffer between the converter output and the analogue controller.

It should be obvious that there is little in either of the above circuits to limit their speeds. In fact, it is generally true that digital-to-analogue conversions are virtually instantaneous and the limitations in the system lie in the often slow speeds of the analogue-to-digital converters.

Analogue-to-digital converters

There is an interesting variety of circuits for the reverse process of analogue-to-digital conversion. A few of these will show how varied these principles are.

The continuous balance ADC

The schematic circuit for this type is shown in Figure 3.8.

This circuit contains a DAC within the loop, normally of the R,2R type. The function of the control circuit is to gate clock pulses to the input of the counter/register section. The latter is just a binary counter with the output buffered by a register which may have parallel or serial output (parallel shown in this case). The state of the control circuit, that is whether it enables or disables the counter input pulses, is determined by the output voltage V_C from the comparator. There are two analogue inputs to the comparator, (i) the voltage V_{IN} to be converted and (ii) the output from the DAC V_p, this being the analogue equivalent of whatever binary value the counter is holding.

Assume that, initially, the counter/register has been cleared by the start convert pulse; the output from the DAC is obviously zero ($V_p = 0$). Assume that a voltage V_{IN} is present that is to be converted. Therefore, $V_{IN} > V_p$ and V_C enables the counter which begins to count up. The output from the DAC rises in a series of small steps as the counter value increases, until there comes a point where $V_p = V_{IN}$ and then just exceeds it. The output of the comparator V_C switches to the opposite polarity and disables the counter, which then stops. The comparator

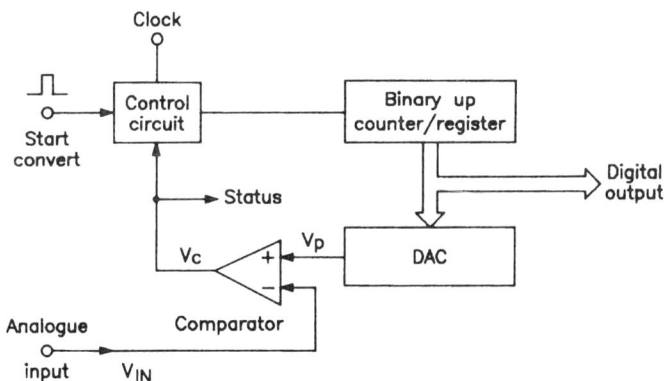

Figure 3.8   The continuous balance type of ADC containing a DAC within the conversion loop

Conversion between digital and analogue

output would also provide the status level output to the computer. The binary value held by the counter/register is the equivalent of the analogue input. The process would repeat on receipt of a new start convert pulse.

This is one of the slower types of converter. The length of the conversion process depends upon the size of the analogue input. For an analogue input voltage equal to the full scale input, the counter has to count up to its limit. An N-bit counter needs $(2^N - 1)$ clock cycles to complete the conversion. An 8-bit counter using a 1 MHz clock would take $(2^8 - 1)$ microseconds, $= 255$ μs to convert the largest input voltage.

The dual-slope ADC

This type of converter, shown in Figure 3.9, is used frequently in digital voltmeters, but is also used in computer input circuits. It does not make use of a DAC in contrast with the previous type.

To start the conversion process the switch S (an electronic one) is set to select the analogue input voltage V_{IN}; at this instant, referred to on the graph as t_0, the output V_0 from the ramp generator is 0 V, as a result of which the comparator output is such as to enable the counter via the control circuit; the latter, also starting from zero, commences an up-counting sequence. While this is happening the output of the ramp generator is a negative voltage slope, this slope being equal to $-K.V_{IN}$. Eventually, the counter will reach its maximum value and, with one more clock pulse, will overflow. This latter event is detected by the control circuit which immediately switches S over to select the $-V_{REF}$ input.

Considering this instant in time when the counter has just overflowed, termed t_1, the ramp voltage has a value that is directly proportional to V_{IN} and the contents of the counter are, of course, zero. From this point in time two events commence simultaneously.

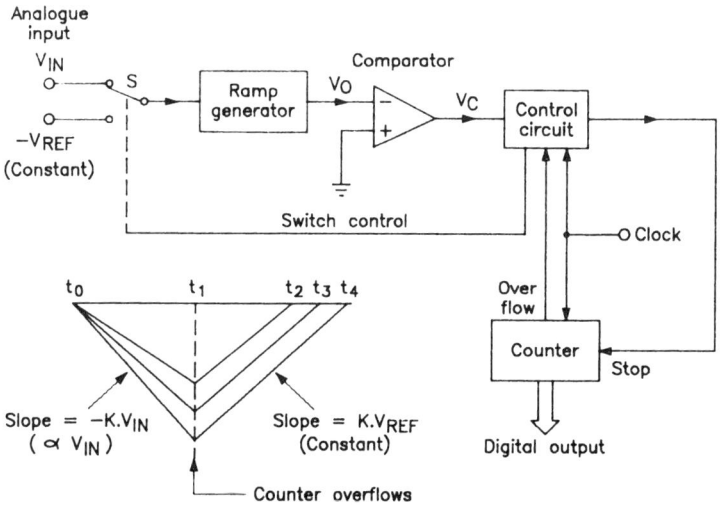

Figure 3.9 The dual-slope ADC, a type also used in DVMs as well as computer inputs

The counter begins to count up again and the output of the ramp generator rises in a positive direction, this time with a slope $K.V_{REF}$. When the ramp generator output reaches 0 V the output of the comparator will invert and cause the control unit to disable the counter. The binary value held by the counter is directly proportional to the time taken for the ramp to return to zero. As the rate which it does so is always constant it follows that the time taken to return to zero depends upon the value of the negative voltage from which it started. Hopefully it is obvious that the latter in turn depends upon the value of V_{IN} the analogue input — hence the relation between the latter and digital output.

This type of converter suffer from the same speed limitation as the continuous balance type for the same reasons.

The successive approximation register ADC

The SAR type of converter represents an ingenious way of obtaining high speed of operation without great complexity. It is an example of another type that makes use of a DAC. The schematic diagram is shown in Figure 3.10.

The principle is quite simple. The converter makes a guess at the value of the binary number required then makes a series of successive approximations, by a totally logical procedure, until it gets it right. A voltage comparator is used to signal when this state has been achieved. An example will make this quite clear. Suppose that the actual binary value required for an 8-bit converter is 10011011 (which of course we don't know at the moment!). The converter always makes the same guess at first, this being the mid-range value of the binary number, in this case 10000000. The comparator tells the SA logic that this guess is too low so the next most significant bit (MSB) is set, giving 11000000. This is clearly too large now, so the second MSB is take out again and the next MSB set, and so on. The sequence looks like this.

Clock pulse SAR contents

	Comparator results	
1	10000000	Too low
2	11000000	Too high
3	10100000	Too high
4	10010000	Too low
5	10011000	Too low
6	10011100	Too high
7	10011010	Too low
8	10011011	Correct

One thing that should be evident immediately is that it only took eight clock pulses to carry out the complete conversion. Some conversions, of course, take less but none will ever exceed N, the number of bits being converted.

To emphasise the matter of speed once more, consider 12-bit converters of the continuous balance and SAR types, both using a 1 MHz clock.

The maximum conversion time for the continuous balance ADC works out at $(2^21^2 – 1)$ µs, which is 4095 µs. By comparison the same conversion carried out by the SAR type would take just N μs = 12 μs!

The hardware for this type of converter can be reduced by using machine code software to run the successive approximation process.

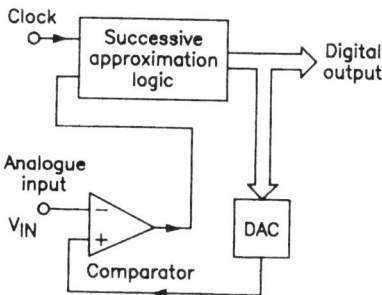

Figure 3.10   The Successive Approximation Register (SAR) type of ADC, offering a considerable speed advantage over the counter-based types

A practical ADC/DAC circuit

The ZN425E IC is a versatile converter chip that can be configured by either a DAC or an ADC. By simple switching it is possible to give it a dual role, though it cannot, of course, be used as both types of converter at the same time. The full circuit arrangement is shown in Figure 3.11. Set-zero controls and full-scale range controls are provided independently for the two halves of the converter.

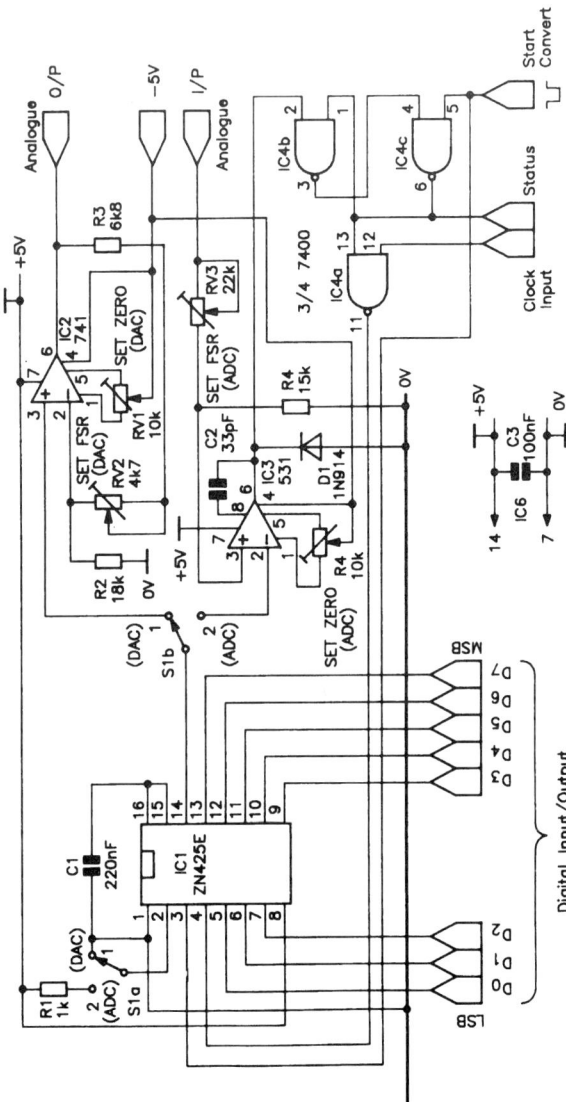

Figure 3.11 A design for a switched ADC/DAC using the ZN425E converter IC

Conversion between digital and analogue

A negative-going start convert pulse is required. The status output is high during conversion and goes low at the end of the conversion process to indicate valid data. A full parts list is included but it is left to the reader to devise a layout.

Parts list for ZN425E switched ADC/DAC

Resistors — all 0.6 W 1% metal film (unless stated)

R1	1 k	(M1K)
R2	18 k	(M18K)
R3	6k8	(M6K8)
R4	15 k	(M15K)
RV1,RV4	10 k vertical encl preset	(UH16S)
RV2	4k7 vertical encl preset	(UH15R)
RV3	22 k vertical encl preset	(UH17T)

Capacitors

C1	220 nF 100 V polylayer	(BX78K)
C2	33 pF 350 V mica	(WX07H)
C3	100 nF 16 V disc ceramic	(YR75S)

Semiconductors

D1	1N914	(QL71N)
IC1	ZN425E-8	(UF38R)
IC2	LM741CN	(QL22Y)
IC3	NE531N	(WQ54J)
IC4	SN7400N	(QX37S)

Miscellaneous

S1	DPDT ultra min toggle	(FH99H)

4   Interrupts

A computer can only perform a useful function by inter-
acting with the *real world* around it, which really means
with its *peripherals*. To perform this process of interaction it must be able to communicate with these
peripherals, either to know when they need attention or
to pass data to them. A common way of establishing this
necessary means of communication is by the method of
interrupts. The interrupt technique forces the computer to respond to a request from a peripheral, at a specific
time, whether because the particular peripheral needs
to pass data to the computer or requires data from it. To
see more clearly how interrupts work and why they are
used, it is worth looking first at an alternative scheme
called *polling*.

Computer interfacing

Software polling

Supposing that a computer is connected to several peripherals, the computer has to know, in some way, when any one of these devices needs service. Polling is a continuous process in which the computer keeps *asking* each peripheral in turn whether it needs attention. Because the speed at which the computer works is so much greater than that of the majority of peripherals, the answer is invariably *no*. In this case, the computer moves on to the next peripheral and repeats the questions, and so on. After interrogating the last peripheral in line the computer goes back to the first one and starts the process again. Such a polling procedure can be executed by a simple segment of machine code that just keeps on looping until a *yes* answer is obtained from one of the peripherals. At this point the computer exits the polling program and jumps to a routine that handles that particular peripheral. The flowchart of Figure 4.1 should make it clear how polling works. There are disadvantages to the polling method, though for some applications it is attractively simple.

For a start, it ties up the computer completely. It is either running the polling program (most of the time) — thus doing nothing very useful — or it is handling a peripheral. There is no chance for the computer to perform any other function during the time that peripherals don't actually need its attention.

Another disadvantage is the relative slowness of response, which is limited by the time taken to run the loop. Supposing that peripheral A requires attention *just*

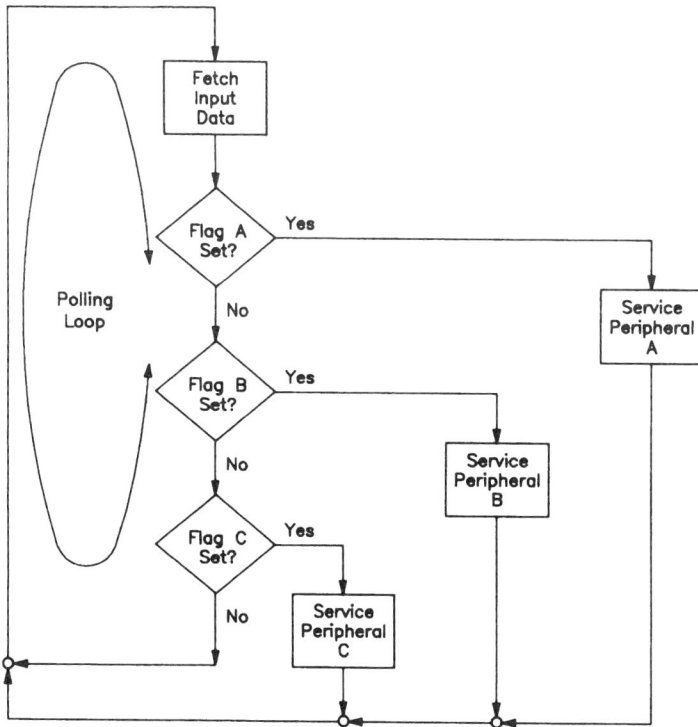

Figure 4.1   Flow chart for polling a computer's peripherals

after it has been checked, it will now have to wait for all the other peripherals to be checked (and perhaps be serviced too) before the loop returns to this peripheral again. Obviously the more peripherals that there are in the queue the slower the possible response.

In the polling method the onus for initiating a routine for handling a peripheral rests squarely on the computer. The peripheral provides a signal (known as a *flag*) when it needs attention but the computer knows nothing of this until that part of the program is reached that checks this specific peripheral. The idea is illustrated in Figure 4.2. Each of the peripheral flags is connected to a single input line on a computer port, and all that the polling program has to do is check each of these lines in turn to establish whether it is *high* or *low*. Assuming that a high state signifies the need for service, on detecting the presence of a high level at one of the port input lines, the program jumps to another area of memory where that peripheral's *handling* routine is stored.

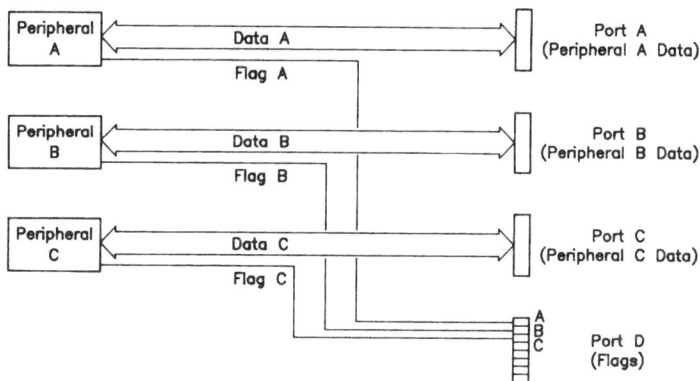

Figure 4.2 Port connections for polling method of servicing peripherals. Port D is checked regularly by the polling programme

By contrast, in the *interrupt* method the responsibility is shifted from the computer to the peripherals themselves. The computer is no longer required to run a polling program and is, in fact, free to pursue some other, perhaps totally unrelated, task. How then do the peripherals force the computer to acknowledge their need for service?

The answer is simply that the peripherals need do no more than they did before, that is to set a flag to an agreed logic level. The difference is in the way the computer reacts to the flag signals.

Figure 4.3 again shows three peripherals, known as A, B and C, connected to a computer. However, now the flag lines do not merely go to a port for status checking but to a *wired-OR* logic circuit as well. The output of this wired-OR arrangement drives a special input to the computer marked $\overline{INT}$. The abbreviation $\overline{INT}$ stands for INTerrupt and the bar over these letters means that the input so designated is *negative-acting*, meaning that it needs to be taken to logic zero in order to initiate the interrupt sequence. This sequence can be summarised as follows:

(a)   the MPU finishes its current instruction,

(b)   various registers, especially the program counter (PC) and status register, have their contents preserved by *pushing* them onto a special area of RAM known as the *stack*,

(c)   the program counter is loaded with the start address of the *interrupt service routine* (ISR) and the peripheral is then handled,

(d)   at the end of the ISR the various registers have their contents retrieved from the stack by a process known as *popping* or *pulling* the stack; the computer is then able to carry on with its original task just as if nothing had happened.

At this stage someone is no doubt wondering how the computer knows which of the peripherals has to be serviced. After all, the interrupt line from the wired-OR logic

Computer interfacing

is common to all peripherals. True, but the flags still go to separate port lines as well, at least in the case of the system shown in Figure 4.3, as there are other methods that can be used.

In this simple scheme the initial response of the computer, after performing steps (a) and (b) in the previous routine, is to go to an address known as the *interrupt vector*. Instead of this address pointing directly to any

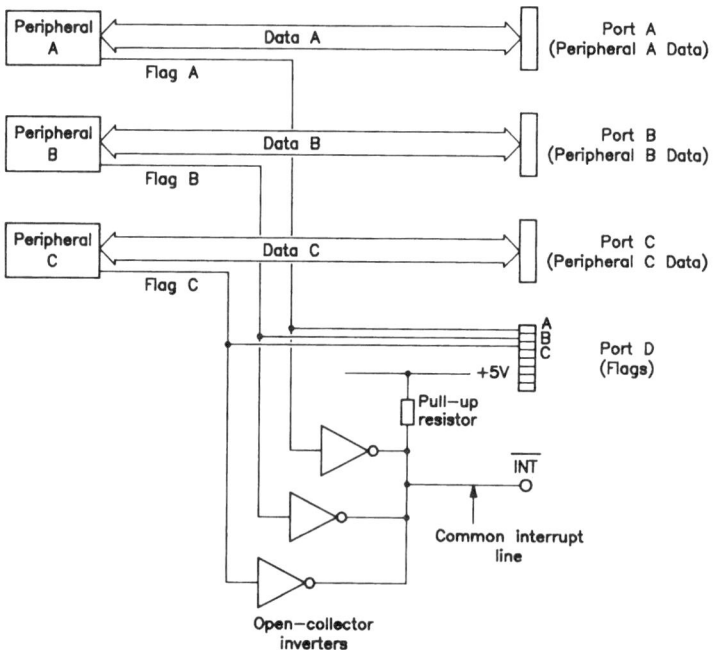

Figure 4.3 Port connections for interrupt servicing of peripherals. Any flag being set (peripheral requiring service) takes the common interrupt line low. Port D is then polled to identify peripheral

specific ISR it is used to run a polling program to check the status of the peripheral flags in turn. On finding a flag set high the program will be directed to the relevant ISR.

In terms of software, this polling procedure is no different from that described previously. The difference, and it is a very important one, is that it doesn't run continuously but is only called when an interrupt request is received. There is no need whatever for the computer to keep on checking the peripherals. Figure 4.4 shows a hypothetical situation where interrupts from the three peripherals of Figure 4.3 arrive quite randomly and are dealt with as they occur. In between times the computer carries on with some other task, known here simply as the *main program*.

Other methods of determining the identity of the interrupting peripheral will be described shortly but first it is worth looking at the idea behind *priorities*.

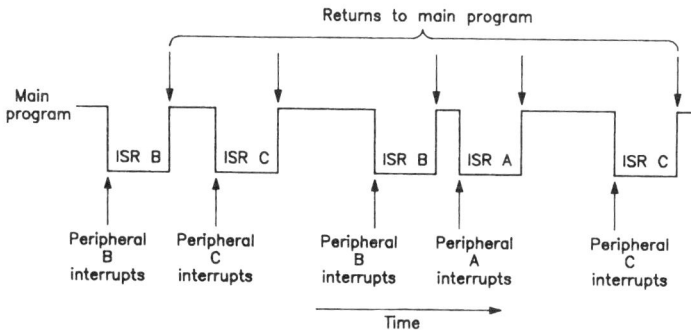

Figure 4.4 Showing how the main programme may be regularly interrrupted by peripherals requiring service. The instants of time are often quite random

Prioritisation of interrupts

In any system controlled by a computer it is likely that some inputs, or peripherals, will have more urgent needs than others. Any potentially hazardous situation, such as over-voltage, loss of power, or excessive temperature rise, must take precedence over the more routine checking of input variables or sending data to a printer, and so on. It must, therefore, be arranged that high priority devices have preference over less vital inputs even to the extent of being able to override their interrupt routines.

The diagram of Figure 4.3 shows only one interrupt connection to the computer, namely the one labelled $\overline{\text{INT}}$. In practice, there is usually a second line frequently known as $\overline{\text{NMI}}$, which stands for non-maskable interrupt. This is the line that is reserved for higher priority inputs. Without prior knowledge, the title *non-maskable interrupt* isn't exactly helpful. It obviously refers to some operation called *masking*, but what exactly is it?

Masking

This is the name given to the setting of a flag, known as the *interrupt mask flag*, in a special register of the microprocessor, in order to prevent a new interrupt routine breaking into one that is already running. Whether this new ISR *should* be able to break into an existing routine depends upon whether its priority is higher or not less than that of the peripheral whose routine is running. When an interrupt is received on the $\overline{\text{INT}}$ input, the microprocessor automatically checks to see if the mask flag is *set* or *clear*. If it is clear, it will allow the interrupt on the $\overline{INT}$ input to continue; if it finds the flag set, it will not allow it to. So, how does this flag get set in the first place? The answer is that it is set by an instruction right at the beginning of an interrupt service routine. This routine could be the $\overline{NMI}$ routine or even one of the $\overline{INT}$ routines.

To take an example of the way in which masking works, suppose that a low priority interrupt is received on the $\overline{INT}$ input and is being serviced. During this time, that is before this ISR has had a chance to finish, another interrupt, also on the $\overline{INT}$ line but of much higher priority, is received but will be ignored because it has been *masked out*. However, if an interrupt is received on the $\overline{NMI}$ line, this will take over immediately since it cannot be masked out. Once the $\overline{NMI}$ routine is finished it will clear the mask bit and the ISR that it interrupted can resume. This process of one ISR breaking into another and *stealing* its time, so to speak, is known as *nesting*.

From the above it can be seen that the difference between the $\overline{NMI}$ and $\overline{INT}$ interrupt procedures is that, when an $\overline{NMI}$ interrupt signal is received it makes no difference whether the mask flag is set or not. This fact is totally ignored and the $\overline{NMI}$ routine immediately breaks into the existing routine and assumes command. In other words, the mask has no effect on the $\overline{NMI}$ input, hence the name, *non-maskable interrupt*.

Now that it can be appreciated that the $\overline{NMI}$ interrupt line will always take precedence, it is then possible to develop the argument a bit further and explain how a system of priority may exist among the peripherals connected to the $\overline{INT}$ input.

Computer interfacing

Suppose that there are three peripherals all wired-OR connected to the $\overline{\text{INT}}$ input line. First it is logical to designate them alphabetically in their priority order so that A has the highest priority and C the lowest. The polling program that is called after the $\overline{\text{INT}}$ line has been pulled low by a peripheral will naturally be written so as to poll them in this order also. This means that, if it happened that two peripherals called for attention at the same time, the higher priority one would be interrogated first and would therefore be serviced first. When its routine had been completed the $\overline{\text{INT}}$ line would still be found to be low, initiating another go at the polling program which would find the other peripheral still waiting for attention; this it would then receive.

However, a moment's thought will show that this has assumed only the simplest of situations. If A is more important than B then if B is already receiving attention, say, at the particular instant that A requires it, it ought to be possible for A to take over from B. This it will do as a matter of course provided that the interrupt mask flag mentioned earlier has not been set by B's interrupt routine. If it has been set, then A would have to wait until B had finished, not really the response that is wanted. On the other hand, if A is receiving attention and B then tries to interrupt, it is essential that it cannot do so. This implies that A's interrupt routine *must* set the interrupt mask flag.

Summing up, the interrupt service routine for A must include the instruction for setting the interrupt mask flag (SEI in 6502 code) at the beginning of the routine, so preventing any further calls on the $\overline{\text{INT}}$ line from having any effect while the routine for A is running. At the end of this ISR the instruction must be included that clears the mask flag (CLI in 6502 code), otherwise all subsequent calls on the $\overline{\text{INT}}$ line will be *locked out* for evermore! The ISR for B, on the other hand, must not include these instructions since it is required to allow A to interrupt if necessary. The problem is that there will be nothing to prevent C from taking over from B either! Obviously such simple systems have their limitations. One solution would be to connect peripheral A to the $\overline{\text{NMI}}$ line, giving it the highest possible priority, and B and C to the $\overline{\text{INT}}$ line with the routine for B written such that it can mask out C.

Daisy chaining of peripherals

Another method of determining which peripheral has initiated an interrupt is the connection of peripherals in a *daisy chain* according to their priorities. This daisy chain is effected by the line called ACK in Figure 4.5. When an interrupt occurs on the $\overline{\text{INT}}$ line the microprocessor goes into its usual sequence of preserving registers and so on, then generates an *interrupt-acknowledge* signal on the ACK line. This is gated to the first peripheral in the chain; if this was the peripheral that initiated the interrupt then it will respond by placing an identification number on the data bus, which the microprocessor will read and use to determine which routine to go to. If the first peripheral was not the one that initiated the interrupt, then the ACK signal will be passed on to B in turn, which has the same choice, either to respond or pass the signal on to C.

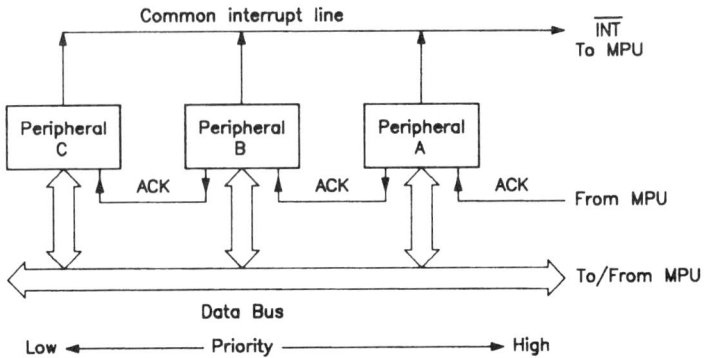

Figure 4.5   Daisy-chaining of three peripherals; highest priority is established for peripheral A

The identification number placed on the data bus by a peripheral can easily be used by the microprocessor to access a look-up table which will, in turn, supply the information regarding which peripheral it is and where in memory its ISR can be found. A quicker response can be obtained if the number supplied by a peripheral points directly to the ISR, so giving immediate access to it rather than by way of the look-up table. This is called a *vectored interrupt*. The Z80 has a facility of this type but because addresses are 16-bits in length and the Z80 only has an 8-bit data bus, only half of this address can be placed on the data bus by the interrupting peripheral. The rest of the address has to be pre-loaded into a special register of the Z80; in practice this is the high byte of the address.

The address formed in this way is not actually the start address of the ISR but merely an *interrupt vector* that points to it. This should be clear from Figure 4.6 in which an address $5C40_H$ is formed which, together with the next address $5C41_H$, provides the actual ISR start address, namely $A120_H$. In other words, the byte 40_H supplied by the peripheral is added to the contents of the I register $(5C_H)$ so as to form the address $5C40_H$. The bytes that comprise the address $A120_H$, already stored at the successive memory addresses, $5C40_H$ and $5C41_H$, then point to the ISR for that particular peripheral.

The Z80's method of performing vectored interrupts (known as mode 2 interrupts) is illustrated in Figure 4.6.

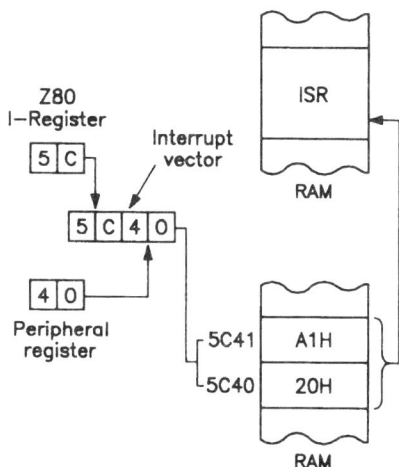

Figure 4.6   The Z80's interrupt mode 2 in which the interrupting peripheral supplies the low byte of the interrupt vector

Computer interfacing

Priority interrupt controllers (PICs)

It is possible to establish interrupt priorities by means of hardware, the *priority interrupt controller* being an example of this. The basic PIC logic is shown in Figure 4.7, which shows how either interrupt lines, number 0 to 7 are handled. Each of these lines is ANDed with the appropriate bit of an 8-bit mask register. The contents of the mask register are determined by the program which loads a byte into this register as required from time to time. Thus, to enable all eight interrupts to be effective, the mask register would be loaded with FFH, that is all *ones*; loading 00H into this register would disable all interrupt lines. To take a further example of this, if 0FH is loaded, the lower four interrupt lines (0–3 inc.) will be enabled while the upper four lines (4–7 inc.) will be disabled.

Being able to mask out interrupts at will within the program leads to a powerful and flexible interrupt handling scheme. Supposing that the interrupt priority scheme is that line 0 has the highest priority and line 7 the lowest then, if an interrupt is received on line 0 it has only to load the mask register with 01H and lines 1–7 inclusive are immediately masked out. Similarly, any other interrupt routine can be written so as to load the required byte into the mask register in order to maintain its priority.

Naturally, lower order priorities will only mask out those below them and not those above, as the latter must be allowed to interrupt them in turn. Thus, if an interrupt is received on line 3 it will mask out lines 4–7 but not lines 0–2.

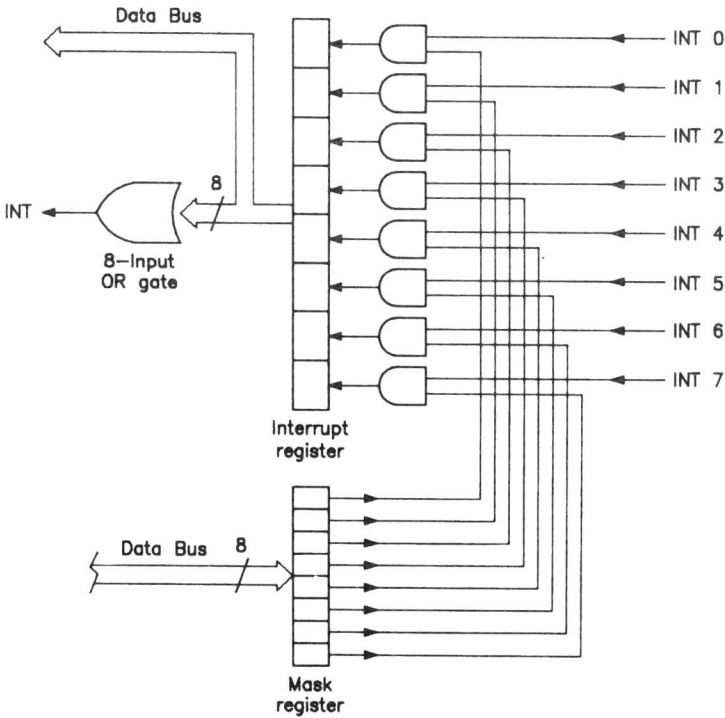

Figure 4.7 PIC (Priority-Interrupt-Controller) logic allowing eight interrupts to be handled; priority can be established by the mask register

The outputs of the AND gates are fed to an 8-bit register which provides two outputs, as follows:

(a)   an interrupt request to initiate the interrupt sequence,

(b)   an 8-bit code which, in effect, identifies which peripheral has initiated the interrupt.

From this 8-bit code a corresponding 3-bit code is generated, using an 8-to-3 line encoder, which can then be compared with the contents of a 3-bit priority register, the contents of which have been specified by the programmer. It is possible to use this result to prevent lower priority interrupts from having any effect. This together with the mask register is very powerful in establishing priorities. The 3-bit code here described is known as a *level vector* and, as a 3-bit code can take up eight different values, in some designs of PIC it is used to access *one-of-eight* 16-bit registers, each of which has previously been loaded with the start address of an interrupt service routine for one of (up to) eight different peripherals. Thus, an immediate branch can be made to the appropriate ISR. This more detailed scheme for a PIC is shown in Figure 4.8.

Direct memory access (DMA)

Interrupts guarantee a very fast response to a call for service from a peripheral but the overall speed is limited by the use of software in the polling and handling routines. Disk drives and monitors may need a faster response than the above method is capable of giving. Since hardware is inherently faster than software, the answer is obvious. Bypass the software aspect with a specialised hardware device. This piece of hardware is known, logically enough, as a *direct memory access controller*, or DMAC. This DMAC is, in effect, a processor designed to perform high speed data transfers between the computer RAM and the peripheral itself. Obviously,

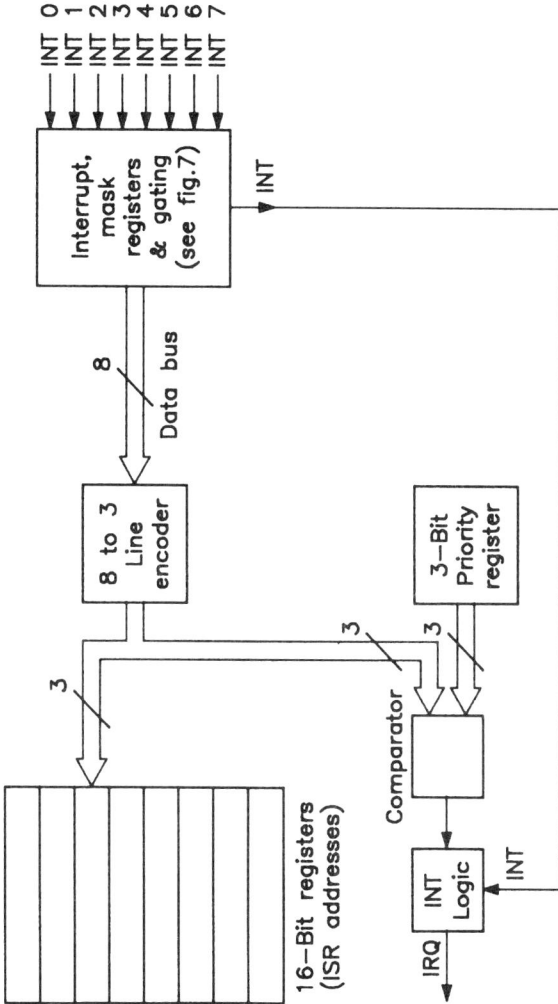

Firgure 4.8 The PIC system of Figure 4.7 expanded to provide direct addressing of ISRs (via the 16-bit register block). Each of eight such registers can be selected by the 3-bit priority register

71

in order to carry out this transfer it needs to have control of both the computer's address and data buses during the time of the transfers. There are various ways of doing this, but the most common method and the easiest to implement is just to suspend the operation of the microprocessor.

A simple concept of a DMAC in operation is shown in Figure 4.9. The operation is as follows.

A significant difference between the operation of a DMAC system and a normal interrupt-driven system is that the interrupt lines from the peripherals no longer go to the microprocessor, but directly to the DMAC. When the latter receives an interrupt request from a peripheral it sends a *hold* signal to the microprocessor which places the latter *in limbo*. However, before it actually enters this suspended state it does three things:

(a)   it completes its current instruction,

(b)   it allows the address and data buses to *float*, that is they go into the third, high impedance, state since they are tri-state devices,

(c)   it sends a *hold acknowledge* (HOLDA) signal to the DMAC to inform it that the buses are now available for its use.

The first action of the DMAC is to place an address on the address bus that specifies the memory address at which the data transfer is to take place. This address is contained in a special 16-bit register in the DMAC; it follows from this that there must be one of these registers for every peripheral connected to the DMAC. The programmer pre-loads these registers as part of the normal

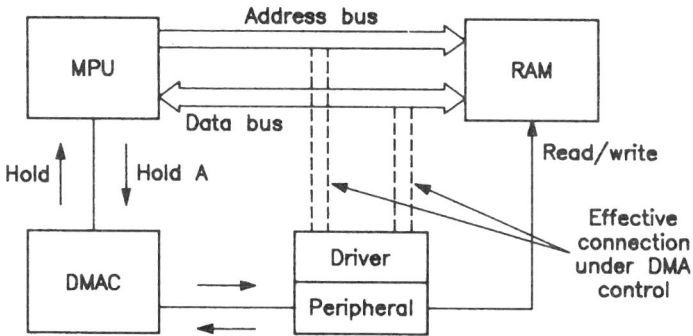

Figure 4.9   Basic scheme for a Direct Memory Access Controller
(DMAC)

boot-up program. The DMAC next sends a *read* or *write*
signal, as appropriate, via the peripheral driver to
memory so that data can then be transferred on the data
bus. A useful facility of the DMAC, allowing it to handle
the transfer of blocks of data, is an automatic sequencing
mechanism controlled by a counter-register, that keeps
track of the number of bytes being transferred in a particular block. As an example of the possible speed of such
devices, the Motorola 6844 is capable of transferring
1 Mbyte of data per second.

Handshaking

Any reference to input-output transfers between a computer and its peripherals invariably includes the rather
colourful term *handshaking*. This is nothing more than a
means of timing the transfer of data to suit both a slow
peripheral and the fast processor. Some idea of how this
is done can be seen from the timing diagrams for output

73

Computer interfacing

transfers (Figure 4.10) and input transfers (Figure 4.11). A feature of hand-shaking is that the input-output port of the computer is connected to the peripheral and not merely by the data bus on which the transfers will be made but also by two hand-shaking lines, known as *ready* and *strobe*.

The output timing diagram of Figure 4.10 shows the logic states, with time, for various lines as follows, starting at the top.

First there is the *write* line which goes low when fresh data is available to be transferred to the peripheral. Next there is the data at the port output itself, the loading of new data into the port register being indicated by the crossing over of the two lines. The next two lines are the handshaking lines themselves and, finally, there is the interrupt line.

The action of outputting the data to a peripheral device implies a *writing* operation from the MPU to the peripheral. Thus, the *write* line will be taken low to initiate the sequence. During the period when the *write* line is low the new data will be loaded into the output port; after a short time interval this data will be considered *valid*, that is the data bus lines will all settle down and it is reasonable to suppose that what is on the data bus is the required byte of data. This will be latched into the output port. The *write* line then goes high again, immediately after which the *ready* line goes high to signal to the peripheral that fresh data is available. When the peripheral has accepted the data, it takes the *strobe* line low to signal the fact to the MPU. At the end of the strobe pulse the output port is cleared and a new interrupt is generated.

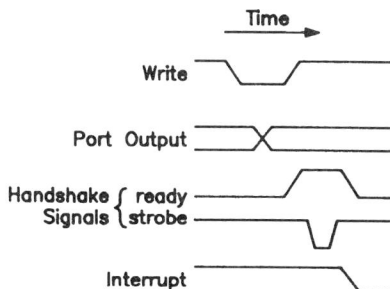

Figure 4.10   Timing for output handshaking

A similar diagram (Figure 4.11) is used to explain the inputting of data from a peripheral. The latter takes the strobe line low and presents data to the computer's input port, which is immediately latched in. At the end of the *strobe* pulse the *ready* line goes low to generate an interrupt; note the $\overline{INT}$ line going low at this time. The *ready* line remains low during the time the MPU is reading the data at the port. The read operation, of course, takes place during the period when the *read* line is low. This occurs quite late in the cycle to allow the data bus to settle down after the new input data has been presented to it. When the *read* line goes high again the *ready* line also goes high, which signals to the peripheral that it is now able to accept more data if required. The subsequent state of the $\overline{INT}$ line will depend upon whether a new interrupt is initiated.

It is hoped that the foregoing explanation of the role of interrupts in interfacing a computer with the real world has aroused some interest in what is a fascinating topic, and perhaps cleared up a few misunderstandings as well.

Computer interfacing

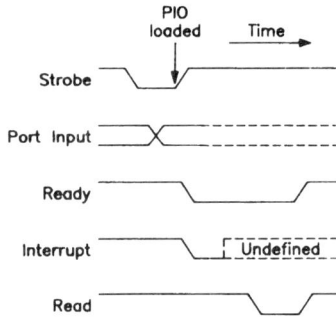

Figure 4.11 Timing for input handshaking

In the next chapter we shall be looking at the serial and parallel transmission of data between the computer and its peripherals and examining some of the variety of interface chips that exist for this purpose.

5   Serial versus parallel

When digital data is transmitted from one device to another, there are two ways it can be done. The bits may all be sent at once, each bit having its own line, this being known as *parallel* transmission. Alternatively, the bits may be sent one at a time, that is one after another in what may be called a *bit stream*. This is known, logically, as *serial* transmission. Obviously with this latter method only a single line (plus a return) is needed.

There are two arguments in favour of sending data serially. The most obvious one is that only the two conductors, one forward and one return, are required, no matter how many data bits there are in the words being transmitted. This has an obvious cost advantage in terms of cabling. The second reason for making use of serial transmission is the ready availability of custom interfaces conforming to accepted standards and thus reducing the risks of physical incompatibilities between the computer and peripherals.

Computer interfacing

Within a computer itself the data is transmitted on a set of parallel conductors, known collectively as the *data bus*. Because all bits of a data word are sent simultaneously, the speed of transmission is very much higher than that for the serial case. The cost is, of course, the greatly increased complexity of the conductor arrangement or, in the case of parallel data transmission external to the computer, the greater cost of cabling.

The nature of serial data

Each bit in a serially transmitted data word is allocated its own *slot in time*. Because the transmission of data in serial form is continuous, but with each data word having a finite length (say, eight bits), it is necessary to *frame* each word with *start* and *stop* bits; this complete set of bits may be referred to as a *data packet*. Usually a serial line will idle in the logic 1 state until a *start* bit signals to the receiving equipment that a data packet is imminent. A typical serial data packet is shown in Figure 5.1. It comprises the *start* bit (at logic 0), followed by eight data bits, ending with the *stop* bit, also a logic 0 level, before returning to the logic 1 idling state. The least significant bit (LSB) is usually sent first. The number of stop bits may variously be 1, 1.5 or 2 bits.

Baud rates

Many serial links are said to be *asynchronous*, which means that a common clock signal is not sent along with data. This implies that the frequencies of the data

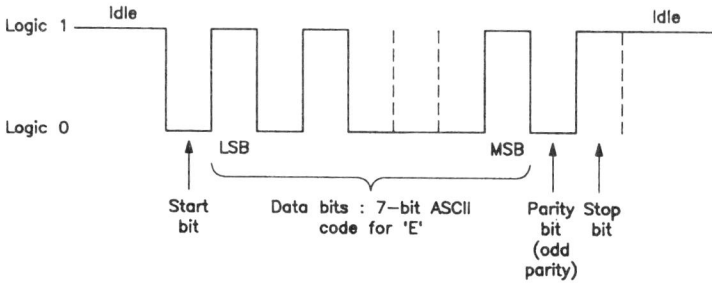

Figure 5.1 A serial *data packet* for the character E (ASCII code = 45$_H$). Odd parity is being used

streams must be identical at both the sending and receiving ends. There are a number of standard frequencies which are actually termed *Baud rates*. these being generally understood to mean the number of bits of data transmitted per second. This is not always strictly true but is close enough for our purposes. The range of Baud rates extends from about 50 to 19,200. Lower speeds are for slow electromechanical devices such as teletypes (if anyone still uses such archaic devices!), while highest speeds are for serial communication with a VDU (*visual display unit*) or with another computer system.

Parity

The continuous transmission of data in the way described above, leads to the ever present possibility of some of the data bits being corrupted, especially if the connecting cables pass through an electrically noisy

79

Computer interfacing

environment. Such corruption of data would mean the inversion of one or more bits; that is a logic 0 becomes a logic 1 or vice-versa. Without some means of detecting such errors the transmitted data would be accepted at the receiving end without question. The simplest technique involves sending an extra bit, known as the *parity bit*, with the data word.

There are two strategies that can be followed in determining the value of a parity bit. They are known as *even parity* and *odd parity* respectively.

In the case of even parity, the number of 1s in the word is counted; if this number is even, the parity bit takes the value 0 — so that the total number 1s in the *data plus parity bit* is even. If, however, the number of 1s in the data word is odd, the parity bit takes the value 1 — again ensuring that there are an even number of 1s in the data packet.

Odd parity follows logically from this, the value of the parity bit being chosen so that there is always an odd number of 1s in the data packet.

The following examples should make this clear:

Data	Parity Bit	
	Even	Odd
01010101	0	1
01101110	1	0
00010011	1	0
11111111	0	1

To conclude the argument, consider the first of the above examples, which is the data word 01010101; this quite clearly contains an even number of 1s, namely four. If this data word is to be sent with even parity then the parity bit must be 0 as the number of 1s in the word is already even. However, if the data word is to be sent with odd parity, then the parity must be a 1 in order that the total number of 1s transmitted is odd (five).

The natural question that follows from the above is, *how is the parity bit used to detect an error?*

It must be said at the beginning that this particular method of error checking using a single parity bit is not very sophisticated and will only detect errors where an odd number of bits becomes corrupted. For example, suppose the data word 10011010 is sent as even parity; the parity bit will be 0. Hence, the data word plus parity bit will look like this:

10011010(data) + 0(parity)

This quite clearly contains an even number of 1s as expected by the receiver and the word will be accepted without question — which is just as well, as it happens to be correct.

Suppose that one bit of the data word becomes corrupted; for example the second bit of the data word changes from 0 to 1. The data word plus parity bit now looks like this:

11011010(data) + 0(parity)

If the 1s are now counted at the receiver, the number will be found to be odd. As the receiver is expecting an even number it will *know* that an error has occurred and will probably ask for the last data word to be re-transmitted. The problem arises, as already mentioned, when an even number of errors occur. For example, if the second and third bits both corrupt to 1s, then the received data will be:

$$11111010(\text{data}) \ + \ 0(\text{parity})$$

The data is incorrect but will be accepted since the number of 1s is even (six), thus conforming to even parity.

Not all serial links employ the parity bit for checking. They either leave it permanently *set* (equals 1) or *clear* (equals 0) or omit it completely.

Data transfer modes: simplex, full-duplex and half-duplex

A *simplex* data link allows the transmission of data in one direction only.

In a *full-duplex* data link, simultaneous transmission of data in both directions is possible.

In a *half-duplex* data link transmission of data can take place in either direction but only on an alternate basis. That is, while one device is transmitting, the other is listening, or vice-versa.

These modes of operation are illustrated in Figure 5.2.

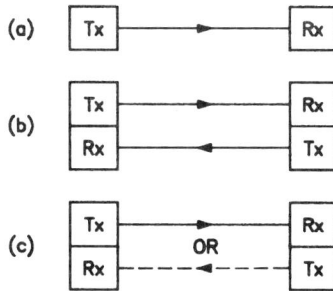

Figure 5.2   (a) Simplex, (b) Duplex and (c) Half-Duplex modes
of data transmission

Parallel-serial conversion

When a serial data link is used, for example between two
computers, the format used within each computer has
to be converted into a serial format in order to use the
single line data path between the devices. There are a
number of dedicated LSI chips available for this function. Some of these will now be discussed, in just
sufficient detail to give an insight into the features that
they incorporate. Some of the mnemonics used may be
familiar:

● UART universal asynchronous receiver transmitter,

● ACIA asynchronous communications interface
adaptor,

● USART universal synchronous/asynchronous receiver transmitter,

● SIO serial input/output device.

The required functions of the above chips can be
summed up as follows, before looking at each chip in
more detail:

83

(a)   use a shift register to perform the serial/parallel or parallel/serial conversion,

(b)   select the required Baud rate, number of data bits and stop bits,

(c)   establish the procedure to be adopted in the event of an error being detected,

(d)   signal the state of the input or output buffer, that is full or empty.

The 6402 UART

This is an industry standard UART, the block diagram of both the transmit and receive circuitry being shown in Figure 5.3. To explain in full how the device is used would occupy the space of most of this chapter, but an attempt to give some idea of its operation *will* be given. This device can be used in a wide range of applications including modems, printers, peripherals and remote data acquisition systems. The CMOS/LSI technology used permits clock frequencies up to 4 MHz with a power consumption of 10 mW or less.

Some aspects of the UART operation are software controlled and some are determined by hardwiring certain of the chip connections. In particular the pins known as CLS2, CLS1, PI, EPE and SBS will be hardwired to a given pattern (from 32 possible patterns) of logic 0s and logic 1s to select the required parameters for number of data bits (5, 6, 7 or 8), parity even, odd or disabled, and the number of stop bits (1, 1.5 or 2). These pins are defined as follows:

● CLS2 and CLS1 (character length selected) allow four binary combinations, corresponding to the four permitted word lengths of 5, 6, 7 or 8 bits specified above,

● PI stands for parity inhibited, this condition being obtained when this pin is high (logic 1),

● EPE stands for even parity enabled; a high level on this pin produces even parity (on transmission) and checks it (on reception),

● SBS has two functions when at logic 1, depending upon the selected word length. For a 5-bit word it produces 1.5 *stop* bits and 2 *stop* bits for other word lengths; however, if SBS is at logic 0 it produces 1 *stop* bit.

There is a relationship between the clock frequency and the Baud rate. The clock frequency chosen should equal *sixteen times the required Baud rate*. Thus, for a Baud rate of 1200, the clock frequency should be equal to 1200 × 16 = 19200 Hz.

Taking the transmit operation first, the parallel data input is loaded into a buffer when the control line TBRL (transmitter buffer load) is taken low. As this data is loaded into the buffer, the fact that the buffer is now in use is indicated to the transmitting device by the output control line TBRE (transmitter buffer empty) going low. The data in the buffer will then be transferred to the transmitter register, the fact of the latter register now being in use being signalled by the second output control line TRE (transmitter register empty) going low. Finally the data in the transmitter register will be passed to the multiplexer, to be supplied with the start, stop and parity bits, before the whole *data packet* is serially

Computer interfacing

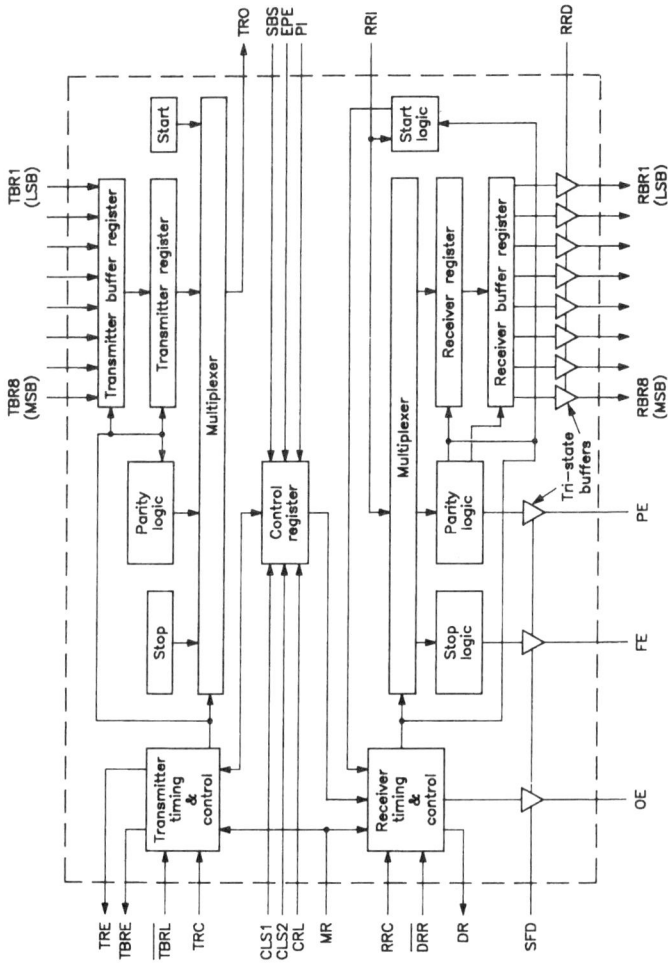

Figure 5.3   Functional block diagram for 6402 UART

86

shifted out onto the line through the TRO pin. All of this is under the action of a clock that, as stated before, will also set the Baud rate for the transmission.

The receive operation is somewhat more complex. The serial data in arrives on RRI. The Baud rate can be different on receive from transmit. There is, therefore, a separate clock (receive clock) and, as before, the clock frequency is 16 times the required Baud rate. The *data packet* received is loaded into the receive register, via the multiplexer, and when the data ready line goes high the parity and format are checked. The data word is then passed to the tri-state output via the receive buffer, where it can be read by the CPU. Once this has been done, the CPU causes the data ready reset to go low, which clears the data ready line. The UART is now able to accept another serial word from the line.

Three types of errors can be detected:

(a)   assuming that parity is in use, a *parity error* will result in the PE (parity error) pin going high, this pin staying high until the next valid character is received,

(b)   a *framing error* occurs when the expected *stop* bit is not received. The FE (framing error) pin will then go high and remain in this state until the next complete character's *stop* bit is received,

(c)   an *overrun error* occurs when a character is transferred to the receiver buffer register before the previous character has been fully read. The corresponding flag for this error is the state of pin OE (overrun error) which goes high to signal the error.

Computer interfacing

The fact that each of the above types of error is *flagged* at an external pin of the 6402 allows the system designer to initiate an error correction procedure, such as the retransmission of the character that caused the error.

There are two ways in which a UART can be used, known as the *unconditional* and *handshaking* modes. In the case of the former, the idea is that any time a character arrives it gets handled. This is simple but carries the reservation that the rate at which such characters arrive must be limited according to the selected Baud rate. In the *handshaking* mode the UART decides, by means of control signals, just when it will handle a character. When a UART is used as a serial interface to a VDU/keyboard the unconditional mode is usually the one employed.

The above much abbreviated explanation does, it is hoped, provide rather more than just an indication of how complex the functions of such a chip may be. There is a dual version of the UART known as a DART (dual asynchronous receiver/transmitter).

The Motorola 6850 ACIA

This device implements the requirements of the standard EIA232 (commonly, though improperly, known as RS232C) serial interface, which is discussed later in this chapter.

The block diagram appears in Figure 5.4. It should be noted that its pins can be defined in *blocks* with specific functions and relations to the microprocessor buses.

The parallel data connection is made through the eight data pins D0–D7 inclusive. Two address bus lines are used for chip selections, these being identified as CS1 and $\overline{\text{CS2}}$; thus one of the selected address lines at logic 1 and one at logic 0 will enable the chip. A third address line determines the logic level on the pin RS and, together with the logic level on the R/$\overline{\text{W}}$ line, provides the four binary combinations that control the addressing of the internal registers, known as *control, status, receive data* and *transmit data*, thus defining the mode of operation of the device.

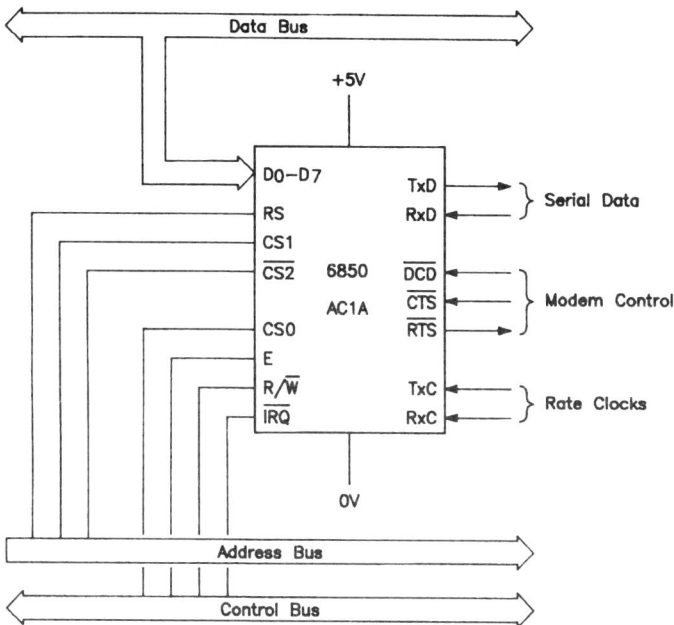

Figure 5.4   Function diagram for 6850 ACIA

Computer interfacing

Serial data in and out are via the pins RxD and TxD; these are TTL compatible signals, but they will invariably require external buffering to interface them to the serial devices. The modem-control lines control the interface in an RS232C modem link. There are two clocks, one for receive and one for transmit and, as explained previously, they may be different for the two directions, thus providing different Baud rates.

The 8251 USART

This device, the block diagram for which appears in Figure 5.5, permits both synchronous and asynchronous serial data transmission. The latter method has been described already and it should be appreciated that the data can be sent at irregular intervals, as the presence of the *start* bit will always identify the beginning of new data. In the synchronous method, data is transmitted quite precisely and the precision of the timing means that the *start* and *stop* bits are not required. Instead the same system clock is supplied to both ends of the serial link by means of a separate track. Every data word has its own predetermined *time slot* into which it fits precisely. Any gaps in the actual transmission are filled by *null words*. Synchronous systems are very fast but at the cost of greater complexity than asynchronous links.

The 8251 has to have *set-up* words sent to it after a reset following power-up. This is a common feature of other *programmable* interfaces and the function of these set-up words is to establish the mode in which the 8251 will operate. Obvious options are: synchronous or asynchronous; read or write; word length and other parameters.

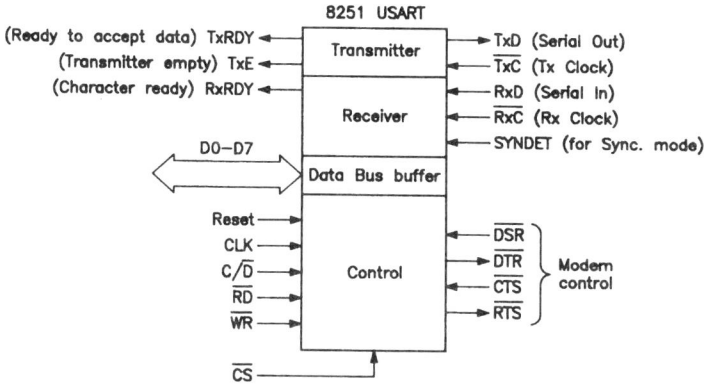

Figure 5.5   8251 USART

The use of the 8251 as an interface between a CPU and the VDU/keyboard is illustrated in Figure 5.6. Another important peripheral interfacing device is shown in this case, namely the CRTC or cathode ray tube controller IC.

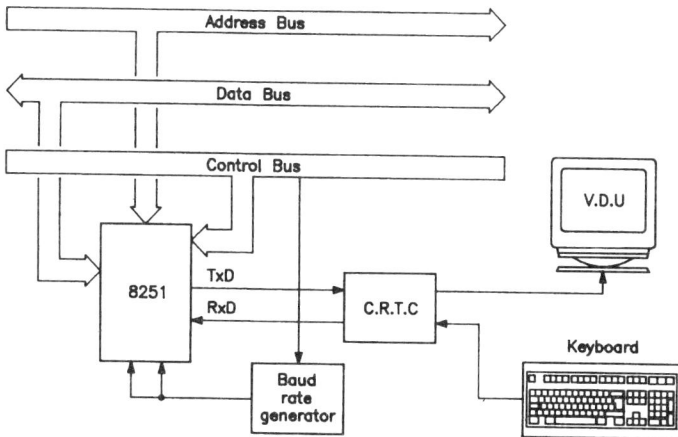

Figure 5.6   Using the 8251 USART to interface a micro to its keyboard and display

91

Computer interfacing

The SIO (serial input/output device)

A serial input/output controller IC is available for the Z80 CPU but it can, in fact, be used with a variety of other microprocessors.

It provides two independent full-duplex channels with separate control and status lines for modem and other devices. The general facilities are much the same as those already specified for the 6402 in respect of the format of the *data packet*, that is the number of data bits, *start* and *stop* bits, parity, etc. Data rates of up to 800 kbit/s are possible with a 4 MHz clock. The usual types of error can be detected and flagged.

The EIA 232 standard serial interface

This interface standard conforms to a quite old Electrical Industries Association (EIA) standard (a US standard, in fact) and is the one often employed with modems and for large-scale computer serial interfacing. It is commonly referred to (incorrectly) as RS232C. The EIA232 signal is bipolar, which simply means that it has two signs, one for logic 0 and another for logic 1. In fact the voltages commonly used are +12 V to represent logic 0 and –12 V to represent logic 1 — see Figure 5.7. This makes it necessary to perform a conversion between the TTL/EIA 232 logic levels in both directions of transmission. The interface devices normally used, and shown in use in Figure 5.8, are the 1488 and 1489 ICs.

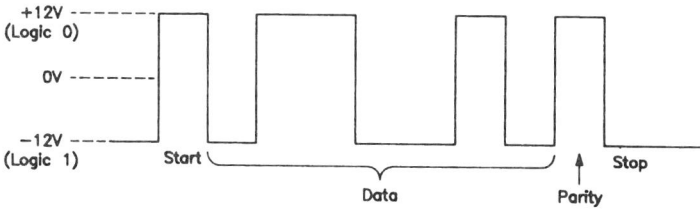

Figure 5.7   RS232C voltage levels

This interface is sometimes referred to under the dual reference EIA232/V.24. The latter is actually a European specification under the auspices of the CCITT and, for practical purposes, may be considered as virtually interchangeable with EIA232.

Data codes

Probably the best known code is the one referred to as ASCII, which stands for *American standard code for information interchange.* Each of the alphanumeric and

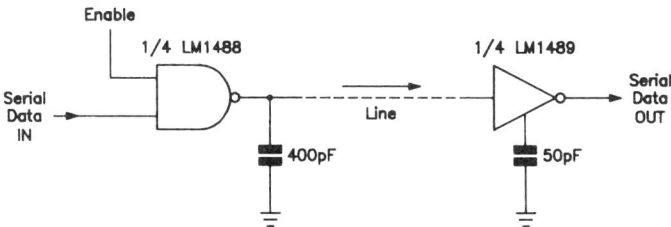

Figure 5.8   Using LM1488/1489 ICs to implement RS232C levels; capacitors minimise ringing and noise

93

Computer interfacing

other symbols on a keyboard have a unique code, which is usually specified in hexadecimal format. In addition to keyboard symbols, various control functions, such as carriage return, line feed — to quote two obvious ones — are similarly encoded. These hexadecimal values range from 00_H to $7F_H$, those from 00_H to 31_H being the control values, the remainder being the true keyboard characters. This code is basically a 7-bit code (thus allowing for 128 different characters to be encoded) with the eighth bit being used for parity. It is interesting to note that bits 6 and 7 of the code define the nature of the ASCII character. All those commencing with 00 are *control* characters; all those commencing with 01 are *punctuation marks*; all those commencing with 10 are *capitals*, while all those commencing with 11 are *lower case* letters. In this way the computer is easily able to ascertain whether a character is, for example, for control merely by noting whether the two most significant bits are both *zeroes*.

An IBM code is the EBCDIC code, which stands for *extended binary coded decimal interchange code*. This is an 8-bit code used in IBM and other compatible computers. There is more than one version of this code.

Modems

Modem is one of those combination words. The *mo* part is an abbreviation of *modulator*, while the *dem* part describes the complementary device, the *demodulator*. The purpose of a modem is to allow data to be transmitted over the public telephone network. Since this normally handles speech frequencies in the range 300–3400 Hz, the logic levels have to be converted into corresponding audio frequency signals. This can be accomplished by a method known as *frequency-shift keying* (FSK) in which logic 0 is represented by one tone (e.g. 2100 Hz) and logic 1 by another (e.g. 1300 Hz). The relationship between a TTL binary signal and its FSK equivalent is shown in Figure 5.9. The function of each component parts of a modem is now obvious.

The modulator section will have a TTL binary input which it will convert into the corresponding pattern of audio frequency tones. Its input will be from the computer and its output will be into the telephone system.

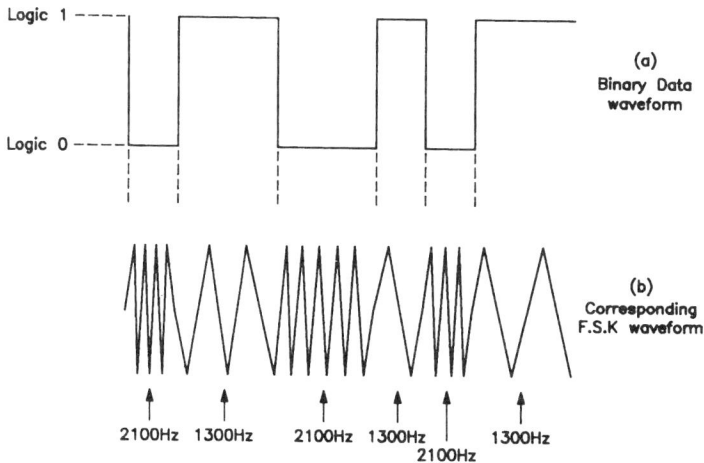

Figure 5.9   In Frequency Shift Keying (FSK), each bit is encoded as one of two alternative audio frequencies, according to its logic level

Computer interfacing

The demodulator will have as its input the audio-frequency FSK signal which it will then convert into the appropriate TTL levels. Its input will be from the telephone line and its output will be into the computer.

As the signals into and out of the line will be in a serial format, and as the computer(s) will require parallel data, the modem will also include the facilities previously described for serial/parallel and parallel/serial conversion. Modems may use the simplex, full-duplex or half-duplex modes of operation. If the two frequencies mentioned above, namely 1300 Hz and 2100 Hz are used for the two logic levels, only half-duplex working is possible. For full-duplex operation there must be four frequencies, one pair for the forward direction of transmission and another pair for the reverse direction.

For example:

1   channel A might use 1850 Hz for logic 0 and 1650 Hz for logic 1, while,

1   channel B might use 1180 Hz for logic 0 and 980 Hz for logic 1.

Acoustic coupling of the modem to a telephone handset has been tried in the past as an inexpensive way of connecting the computer into the public telephone network, but it brings its own special crop of problems. It is more usual to make a direct electrical connection instead with suitable isolation arranged to prevent inadvertent connection of mains voltages to the telephone lines, plus various other sophistications. As already mentioned, the usual form of interface employed corresponds to the EIA 232/V.24 standards.

6   The computer keyboard and VDU

There is a variety of devices which allow a user to input data directly into the computer. Of these the keyboard is certainly the most familiar and commonly used. Other devices, such as the mouse, trackerball, graphics tablet and so on, act merely as sometime alternatives to the keyboard and not as complete replacements.

The complexity of a keyboard may vary between the spartan simplicity of a hexadecimal keypad (providing nothing more than the hexadecimal digits 0–9, A–F, plus a few control functions) and the luxury of a full typewriter-style keyboard. The latter may well have been expanded even further by the addition of a set of function keys and special control keys, according to the exact nature of the computer with which it associated.

Computer interfacing

Keyboard problems — rollover and lockout

It is virtually impossible to type without, at sometime pressing two or more keys in such rapid succession that they are almost instantaneous. Without some effective method of guarding against the consequences, this could result in the wrong codes being generated. There are three available techniques to prevent such errors. They are known as, *n-key lockout*, *2-key rollover* and *n-key rollover.*

The simplest and cheapest is n-key lockout. In this method, the first key down generates a pulse known as a *keyboard strobe.* This causes the code for that key to be fetched. During this period any other keys down are totally ignored. The obvious disadvantage is that there are likely to be a number of missed keys, since one key must be fully released before the next is depressed.

In 2-key rollover, on the other hand, any key down will generate a keyboard strobe; a delay is then introduced for any further keyboard strobe until the first key has been released. The second key down will then be accepted. This is a method that works well as long as no more than two keys are down at a time.

The n-key rollover method is generally considered to be a luxury without much merit because of the considerable degree of complexity of hardware it requires. Its operational feature is that any number of keys down will immediately generate the required codes for those keys, in the sequence of the keys pressed.

The computer keyboard and VDU

The keyboard matrix

The electrical pattern beneath a keyboard is usually in the form of a matrix of conductors, consisting of *m columns* by *n rows*: this matrix is not necessarily physically similar to the keyboard layout. The keyswitches are each wired across a column-row intersection. Thus, an 8-column by 8-row matrix will accommodate up to 64 keyswitches. The use of such a matrix offers the possibility of a system in which the columns (say) are the input lines and the rows are the output lines. If no key is held down, then there is no conducting path between any of the columns and any of the rows. Naturally, a key down will then provide such a path. Every path is unique and is, therefore, potentially capable of identifying which key is down at any time. An obvious problem in this respect is that an input on any of the *n rows* could have originated from any one of the *m columns*. Some means must be found of making each column identify itself when a keypress occurs. One method of doing this, called *scanning the keyboard*, will be described shortly.

Non-encoded and encoded keyboards

All keyboards fall into one of the two above categories, though most will occupy the latter one. In a non-encoded keyboard, the process of scanning, and subsequent key identification, is largely software orientated. In the encoded type of keyboard, the accent is on the hardware.

Taking the example of a simple 16-key keypad, this would be organised as a 4×4 matrix, as shown in Figure 6.1. To scan it, a 4-bit binary pattern, in which a single logic 1 is constantly circulated through the pattern, is applied to the four column wires. Thus, each of these wires is energised in turn by the logic 1, all others at this instant being at logic 0. Such a binary pattern is easily generated by software, and sent to the column wires via four lines of the computer's input/output port, dedicated for this purpose. The successive states of the four column lines will be, starting with 1000, next 0100, 0010, 0001, then 1000 and so on, the sequence repeating indefinitely. The result of scanning the columns in this way is that, when a key is down in the column which is at logic 1, this logic level will be passed out through the appropriate row line. As the scanning rate greatly exceeds the speed at which even the fastest typist can press successive keys, there is no chance of the keypress being missed.

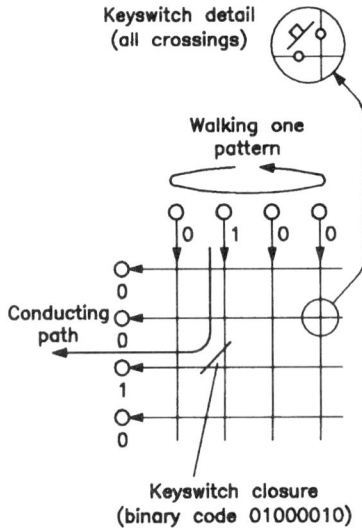

Figure 6.1 A 16-key keypad with the columns scanned by a 'walking one' pattern

The computer keyboard and VDU

A moment's thought should show that the *key down* has now been effectively identified. Each of the four column patterns listed above can form a possible combination with each of the four rows in which a keypress can occur. This gives the sixteen unique combinations or codes required for the sixteen keys of the keypad. For example, the code 01000010 (reading columns top to bottom; rows left to right) identifies the key that is in the second column in, third row down, as the one that has been pressed. The software must be written so as to include the following features:

(a) to identify when there is a 1 in the second half of the number (the row bits), as a 1 only exists here when a key is down,

(b) to read the complete number so as to check it against a table of such numbers, held in memory, and thus identify the key pressed.

The program must continually be checking for the 1 on the row lines, as described, and this will correspond to the action of the *keyboard strobe pulse* (see later). On finding the logic 1, the program will then jump to a routine that checks the whole number as described in (b) above.

This idea, in which a logic 1 is used to scan the keyboard lines, is known by the rather picturesque title of *a walking one's decode*. In the alternative approach the *walking one* pattern is generated by a piece of hardware, of which the ring counter (with one stage only set) is a prime example. The ring counter is a type of shift register that consists of a string of clocked D-type flip-flops in which the Q output of the final flip-flop is fed back to the D input of the first. Thus, the data is circulated continuously at the clock rate. As only one flip-flop was set, the data will consist of a single 1 and $n - 1$ zeros (when n is the number of stages of the register).

In the case of a 64-key keyboard, the matrix would be organised as an 8×8 layout — eight columns and eight rows. A scan pattern on the columns and rows could be generated by an 8-stage ring counter for the columns and a *1 of 8 selector* for the rows. However, there is an important point here; either the columns ring counter must be clocked at eight times the rate at which the 1 of 8 selector is clocked, or vice-versa. For example, if the rows are clocked at a frequency *f,* then the columns could be clocked at a frequency *8f.* In this way, a row is held selected during the time that the *walking one* energises each column in turn. Then the next row is selected while all the columns are energised again. In this way all keys are scanned in a logical pattern, row by row. A scheme along these lines is shown in Figure 6.2.

The codes for each of the keys in the matrix are contained in a ROM, which is also organised on an 8 x 8 basis. It is virtually a *carbon copy* of the key matrix in this example; the spatial arrangement of memory locations where the key codes are stored mimicking the layout of the matrix. The coincidence of row and column obtained by a keypress generates a logic 1 on similar lines addressing the ROM locations. As a result, an 8-bit code is output from the ROM, from where it will be passed to the CPU.

ASCII code

In theory a variety of coding systems could be devised to encode the keys of the keyboard. In practice there is

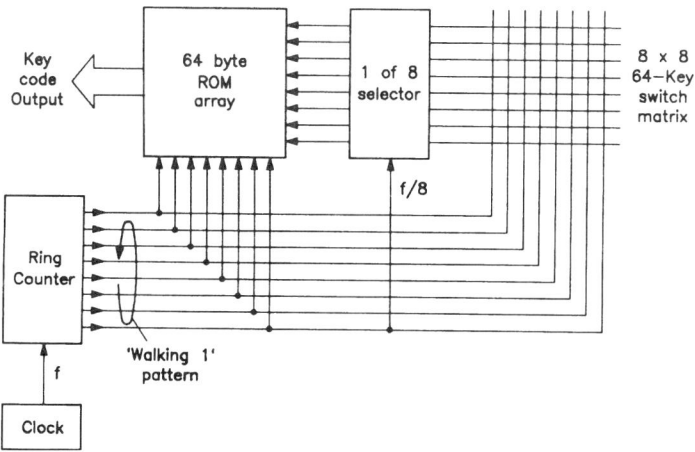

Figure 6.2   A 64-key keyboard, with the columns hardware scanned by a ring counter; the key codes are held in a special ROM, which is addressed by the coincidence of row/column key closure

one code in general use and a few others that may be encountered in certain circumstances. The code most often used is termed ASCII, which is an acronym for *American standard code for information interchange*. This is generally regarded as a 7-bit binary code, though an 8-bit version also exists. This code is shown in Table 6.1, and it will be seen that all alpha-numeric characters, punctuation marks, other special characters and also a variety of control characters, are to be found in this code. Each is represented by a two-digit hexadecimal number.

A scanning encoder is shown in Figure 6.3 in which the ASCII codes are generated directly by the keyboard matrix plus some extra logic. There is no ROM to store the codes in this case. Also included are two keys, of much

Computer interfacing

Control		Numeric	Upper case	Lower case	Special
00	Null	30 0	41 A	61 a	20 Space
01	Start of heading	31 1	42 B	62 b	21 !
02	Start of text	32 2	43 C	63 c	22 "
03	End of text	33 3	44 D	64 d	23 #
04	End of transmission	34 4	45 E	65 e	24 $
05	Enquiry	35 5	46 F	66 f	25 %
06	Acknowledge	36 6	47 G	67 g	26 &
07	Bell	37 7	48 H	68 h	27 '
08	Backspace	38 8	49 I	69 i	28 (
09	Horizontal tabulation	39 9	4A J	6A j	29)
0A	Line feed		4B K	6B k	2A *
0B	Vertical tabulation		4C L	6C l	2B +
0C	Form feed		4D M	6D m	2C ,
0D	Carriage return		4E N	6E n	2D -
0E	Shift out		4F O	6F o	2E .
0F	Shift in		50 P	70 p	2F /
10	Data link escape		51 Q	71 q	3A :
11	X-on		52 R	72 r	3B ;
12	Tape		53 S	73 s	3C <
13	X-off		54 T	74 t	3D =
14	Device control 4		55 U	75 u	3E >
15	Negative acknowledge		56 V	76 v	3F ?
16	Synchronous idle		57 W	77 w	40 @
17	End of transmission block		58 X	78 x	5B [
18	Cancel		59 Y	79 y	5C \
19	End of medium		5A Z	7A z	5D]
1A	Substitute				5E ^ or ↑
1B	Escape				5F _ or ←
1C	File separator				60
1D	Group separator				7B {
1E	Record separator				7C
1F	Unit separator				7D }
7F	Rub out or delete				7E ~

Table 6.1   The standard ASCII keyboard codes and their hexadecimal values

The computer keyboard and VDU

value on computer keyboards in particular, namely *shift* and *control*. Each of these used in conjunction with other keys allows many more functions to be generated. The *shift* key, of course, produces the capital letters plus those punctuation marks in superior positions on the key caps; it also has some special uses in application programs such as word-processors, as does the *control* key.

In this design an 8×8 keyboard matrix is addressed by the outputs of a 1-of-8 decoder (column scan) and a 1-of-8 data selector (row scan). The two logic circuits have

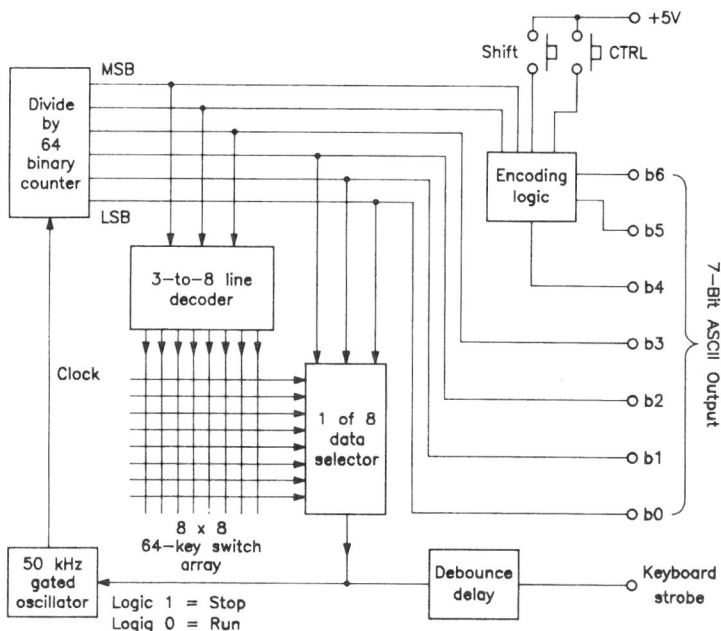

Figure 6.3 A keyboard in which the ASCII codes are generated directly by the hardware. Two-key rollover is inherent in such a design

105

3-bit binary inputs provided from the 6-bit output of a *scale of 64* binary counter. The decoder takes the most significant three bits of this counter and uses them to generate a *walking one* pattern on the columns. The data selector takes the least significant three bits and scans the rows for the appearance of a logic 1 (representing a keypress) at *one-eighth* of the frequency that the columns are being scanned. This essential frequency relationship between the two scanning rates arises naturally because of the binary division between successive stages of the binary counter. That is, the overall division ratio of the counter is 64, this being achieved because the circuit is, in effect, one *divide-by-8* counter (that is for the columns) followed by a second identical counter (that is used for the rows).

The logic 1 output from the data selector, obtained when a keypress occurs, performs two functions. After a *debounce delay*, it is used as the *keyboard strobe* to inform the CPU that a keypress has been detected and that data is available. It is also used to inhibit the 50 kHz gated clock oscillator, thus preventing further keypresses from having effect while the current keyboard data is being accepted by the CPU, so providing inherent two-key rollover. This inhibit of course lasts only for as long as the key is down since, on the key being allowed to return to the normal position, the logic level on the inhibit line will return to logic 0 and the oscillator will restart. Relative to the speed that a human operator will depress and raise a key, the response of the CPU in detecting the keypress and accepting the data is extremely fast. There is therefore, little chance of the keypress going undetected.

The computer keyboard and VDU

A standard keyboard encoder

The General Instrument AY-5-2376 keyboard encoder is an IC that has achieved some degree of acceptance as a standard basis for the design of computer keyboards. This LSI device is a 2376-bit read only memory that has the capability to encode single-pole, single-throw keyboard closures into a 9-bit code. The data and strobe output are directly compatible with TTL or MOS logic, thus obviating the need for any special interfacing. Among the features found on it are:

(a)   provision of external control for output polarity selection,

(b)   provision of external control for selection of odd or even parity,

(c)   two-key rollover operation,

(d)   n-key lockout,

(e)   externally controlled key *debounce* network,

(f)   internal oscillator circuit.

As stated already, the 2376 contains a ROM, which is organised as a 264×9-bit memory arranged into three 88-word by 9-bit groups. The *shift* and *control* keys permit switching between the three 88-word groups. The 88 words themselves are addressed by the ring counters so that the actual address in the ROM that is accessed for data is formed from the 11×8 ring counter matrix with or without either the *shift* or *control* keys. This, in theory, provides for a total of 264 keyboard characters, but this is rarely taken up.

Computer interfacing

The maximum of 88 keys that can be used with this IC should be wired on an 11 × 8 key matrix with a single-pole, single-throw switch at each crossing. In the *standby* state, with no keys pressed, the ring counters sequentially address the ROM locations but the absence of a keyboard strobe pulse indicates that there is no valid data available.

An application circuit for the 2376 keyboard encoder is shown in Figure 6.4. When a key is depressed, a path is completed between one of the outputs of the 8-stage ring counter (X0–X7) and one of the inputs of the 11-bit comparator (Y0–Y10). Shortly after, the comparator sends a signal to the clock control and strobe output (via the delay network). The clock control stops the clocks to the ring counters and the data on the data outputs is now valid, as indicated by the strobe output. This data remains stable until the key is released.

Other forms of input device

Although the keyboard is the most commonly employed device for direct communication with the computer, there are others that have, under certain circumstances, particular advantages. It is worth considering the limitations of the keyboard in order to better understand why there might be a need for alternatives at all, and what is actually being considered as alternative means for a human operator to input data, in ways that may be more ergonomically sound.

For inputting text the keyboard is obviously ideal; all that is needed is the ability to type with reasonable ease.

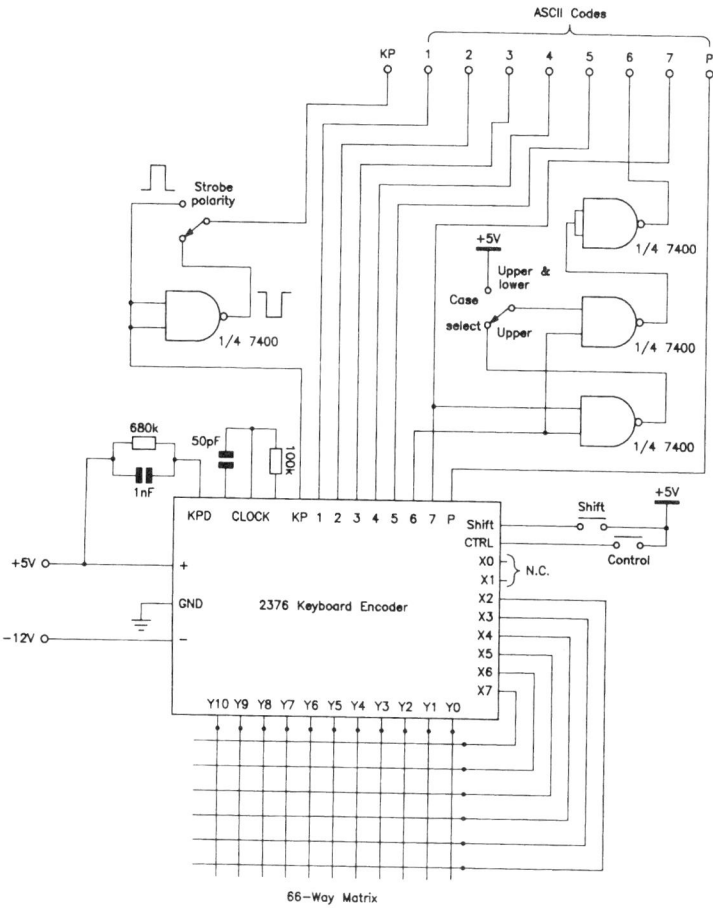

Figure 6.4   An application circuit for the 2376 keyboard encoder IC. Facilities include choice of polarity for the keyboard strobe, and choice of case: upper plus lower or upper only

Probably most computer keyboard users have never had any formal typing training, but this rarely seems to matter. What the keyboard doesn't do quite so well is to allow

Computer interfacing

control of what might loosely be termed, *non-textual* matter, such as graphics. The implications here are broad. Graphical on-screen data can be found in the most serious of software as well as in the most trivial. For positional control using the keyboard, the *cursor keys* are frequently used. These give fairly obvious up/down, left/right control but the feel obtained is not totally natural. For movements intermediate between the four directions stated, and for sudden reversals of direction, they are positively clumsy. An excellent direct replacement is the *joystick* — hardly an unfamiliar object to many computer users. Connected to the computer through a standard 9-pin *D* connector, the switch closures caused by the side-to-side, to-and-fro, movements are read by software. A key closure is a true binary event; either it is closed or it isn't. The two possibilities represent the binary values, 1 and 0. There is no middle choice. Not only can the software identify the four basic joystick movements; it can also tell four intermediate states, when two switches are closed at the same time: NE; SE; SW and NW. A *fire button* is also a usual feature of a joystick and gives the software another switch closure to detect. Naturally, if the coincidence of this event together with the other switch closures is detected, this gives quite a few possibilities. The joystick, simple as it is, starts to look quite sophisticated. Ergonomically it is excellent. The handle sits comfortably in the hand with either a finger or thumb over a fire button. The natural feel is able to provide rapid, co-ordinated screen control, as needed in arcade and simulator games, but it can also be used in more serious applications. While there are some graphics software packages of the art/drawing variety that use the joystick, most serious software looks elsewhere for a suitable input device. Enter the mouse!

The computer keyboard and VDU

The mouse

This colourfully named data entry device is an increasingly popular method of screen manipulation. It is not confined solely to *arty* applications, but is also widely used as a means of communication in what is usually referred to as a *WIMP* environment (W = windows; I = icons; M = mouse and P = pointers). In this type of situation, of which the well known Windows and GEM (Graphics Environment Manager) are examples, the keyboard is largely ignored. Instead of typing in textual commands to obtain the required response, the mouse is used to move a pointer around the screen, which may contain text in windows and/or pictorial symbols (the icons). By positioning the pointer over appropriate text or icon and *clicking* on the mouse, the required commands are ordered and executed.

A mouse is a highly developed form of joystick, in essence, though not necessarily a very close relation. In some types of mouse the switches of the joystick have been replaced by a pair of orthogonally mounted (that is, mutually at right angles) potentiometers. This arrangement is similar to the joystick control in a radio control transmitter. However, instead of using a stick to move the potentiometers, a ball on the underside of the mouse body is rolled on a suitable surface and this motion is transmitted to them.

Because the signals developed by potentiometers are analogue, an interface must provide analogue-to-digital conversion. Most mice, because of this, use orthogonally mounted optical digital shaft encoders (see page 31) which give true digital indication of the mouse position directly interfacing with the computer.

Computer interfacing

The mouse is small and compact and sits comfortably in one's hand, the two or three switches, also found as a feature of the mouse, being positioned where the fingers fall on them naturally when required. A mouse often operates on a special non-slip *mouse mat* and may even be kept in a *mouse house* when not in use — though, perhaps mouse *garage* would be more appropriate! The mouse's *tail* is, of course, merely the cable that connects the mouse to its special interface! The latter may be external to the computer (where the mouse is not an integral feature of that particular machine) or be housed within the computer, where the mouse is supplied as part of the ensemble. The extra complications of the mouse are worth it. Positional control is smooth and positive, and the mouse can also detect, not merely position in terms of its X-Y co-ordinates, but also *rate of change* of position.

The trackerball

This device is, in effect, a mouse *upside-down*! It has the principal characteristics of the mouse without, unfortunately, the latter's natural feel. The mouse allows positional control — the user's hand cupped downwards over the mouse body can lightly move it however required while leaving the user's fingers free to *click on* at any moment. Some people complain that the only way to perform the same combined operations with a trackerball is by using both hands — one hand spins the ball, while the other hovers over the buttons ready to depress one — but (depending on the trackerball) it *is* possible to use one satisfactorily. The figure for the total number of trackerball users is, however, probably quite low.

The light pen

This device, made in the shape of a pen, is used to point at some object of interest on the VDU screen. Because of the way in which the display is produced (the basic electron beam scanning process), there is a known time relation between the commencement of the scan and the position of the point at which the light pen is aimed. This makes it possible to identify what exactly the light pen is looking at. Provision for the light pen facility may be included in a special cathode ray tube controller IC, such as the 6845.

As an example of its use, suppose that a number of options in a program are represented on screen by boxes that each contain a different symbol. If now the tip of the light pen is placed over a particular box, each time this area is scanned by the electron beam, there will be an output from the pen's optical sensor. Usually a key on the keyboard is pressed at this time and the coincidence of pen position and keypress initiates the required event.

The optical sensor, which may be a photo-cell, photo-diode or phototransistor, may either be within the tip of the pen itself or in the pen body. In the latter case, an optical fibre link is used to transmit the light picked up from pen tip to sensor.

Another application of the light pen, apart from the above which could just as easily be performed by a mouse and pointer, is in the actual process of on-screen drawing. Naturally, the light pen is still working in the same way as previously described, but now the *event* that it is initiating is producing a line, shape or pattern on the screen in real time. Drawing in this way may be *fun* for a while but it is much more natural to draw on a horizontal surface than on a vertical one. The novelty wears off, even with young children.

The writing or graphics tablet

This is a square or rectangular tablet over which a stylus is moved to give continuous X-Y co-ordinate information. In any position on the tablet the press of a button loads the co-ordinates for that position into the computer. The movements on the *digitiser tablet*, as it may also be called, are reproduced on screen. Apart from originating new artwork in this way, the tablet and stylus can also be used to trace an existing drawing. The latter is placed in position on the tablet and the stylus run over the details of the drawing. The copy is *echoed* to the screen. It can later be printed out. In case it might be thought that photocopying would be easier, it should be pointed out that such copying can be as selective as one wishes, allowing detail changes to be made. All of the data can be stored on disk, to be called up for further changes at a later date.

There are several types of writing tablet:

(a)   the *wire mesh* type consists of a mesh of fine X-wires separated from a similar mesh of Y-wires by a thin mylar film. These are buried just below the surface. Gray-coded patterns of pulses are applied to the wires. At any given X-Y position there is a unique Gray-coded value, which the stylus can pick up by capacitive linking,

(b)   there is a type that works on the *voltage gradient* across the tablet from one side to another, whether across from side-to-side or top-to-bottom. A resistive material, such as *tedeltos paper*, may be used and the stylus picks up the voltages at any point on the surface. These uniquely identify the X-Y co-ordinates of that point,

(c)   in the *pressure sensitive* tablet, the stylus pressure causes a direct electrical connection between closely spaced X and Y wires. All such intersections are, of course, unique and so identify the position of the stylus.

There are other types of tablet that use either acoustic waves or modulated high frequency signals. A major use of writing tablets is as part of a computer aided design (CAD) system.

Up to now, all the devices described in this article have been designed for the express purpose of putting data *into* the computer. They, therefore, represent only one link in the chain of interactive communication between the human operator and the computer. The output *from* the computer can be in several forms, either temporary or permanent. The latter is usually termed *hard copy* and is what is obtained from printers, plotters, and so on. The temporary output, as it is termed here, is the direct, real-time feedback to the user, presented on the screen of a VDU. The latter device is what this final part of the chapter is about. Printers and similar output devices will be the subject of a later chapter in this book.

The VDU (visual display unit)

Sometimes referred to as a *monitor*, the VDU is the standard display device for the majority of computer users. Not only does it *echo back* the user input from the keyboard thus completing an interactive feedback loop, but may be used to provide information, act as a graphics working area, design screen and so on.

There are several divergent paths to follow in the search for a suitable monitor. It is possible to either buy a purpose-built design (the more usual choice for most users nowadays), or a TV receiver can be used if the video output from the computer (plus synchronising information) is made to modulate a compatible UHF signal. The latter choice will not be considered further. The price of a computer monitor varies greatly, being influenced by choice of colour or monochrome and degree of resolution chiefly. What is actually available and the likely cost can easily be learned by scanning the advertisement pages of the computing press. What is of more interest here are the basic principles employed for putting the *dot matrix* pattern for text or graphics on the screen where required.

First it is necessary to dispel any doubts as to what the term *dot matrix* means. See Figure 6.5, which shows the

116

form of several dot matrix characters. A look through
any dot-matrix printer manual will reveal similarly constructed characters. But whereas a dot matrix printer
produces a character by electromagnetically firing fine
pins at an inked ribbon, the VDU screen achieves a similar objective by turning on or off a minute dot of light
formed by the bombardment of the screen phosphor by
an electron beam (the so-called *cathode ray*). In a character matrix consisting of 8 rows by 8 columns, there
are clearly 64 dots that can be energised, or not, in a
great variety of combinations to give an equal number
of possible characters. Often the actual character only
occupies a matrix of 7 rows by 5 columns, but the larger
8 × 8 matrix is needed to allow space between characters.

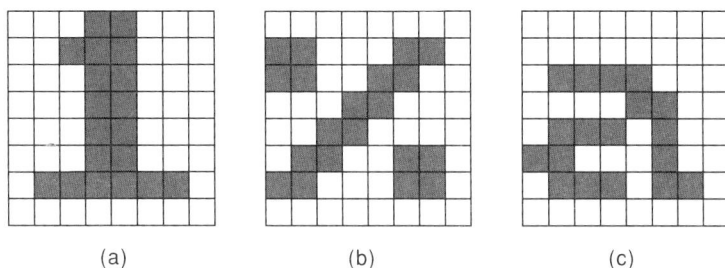

(a)   (b)   (c)

Figure 6.5   Examples of VDU 8 × 8 dot matrix characters: (a)
number 1; (b) percent symbol and (c) lower case a

This basic idea is quite straightforward and probably
familiar to most readers. What is somewhat more eru-

dite is the way in which the dot matrix characters are actually sent to the screen in the required positions, and then moved about at will.

A memory-mapped VDU

To understand what the above term means, it is necessary to consider in more detail the layout of a VDU screen. The smallest element that can be displayed is the dot mentioned previously, which has the special name *pixel*, a contraction of the term picture element. The resolution of a screen may be expressed in pixels, e.g. 640 (horizontal) by 200 (vertical). Another way is to talk about so many columns by so many lines of text, e.g. 80 columns by 25 lines. A moment's thought shows that these are just two alternative ways of expressing exactly the same thing.

Remembering that a single character occupies 8 pixels by 8 pixels, a screen that is 640×200 can hold 640/8 = 80 (columns) by 200/8 = 25 (lines) of text, as before.

The best resolution is obtained by making use of the basic pixel size as the finest detail of the picture (talking about graphics rather than text). This resolution can be *degraded* by combining pixels, in a block of four for example, so giving rise to what are called *chunky graphics*.

Returning to the 640×200 screen, the total number of pixels will be equal to 640×200 = 128,000. As a pixel is a binary value (it is either *on* or *off*) it can be thought of as a *bit* of data in the usual way. Eight pixels will, therefore, occupy one byte and a whole 8 × 8 character will occupy eight bytes. The total number of bytes for a full screen of 128,000 pixels will be 128,000/8 = 16,000 bytes. This is almost 16 K of data; 16 K actually equals 16,384 bytes in computer terms. Ignoring the odd 384 bytes, one could allow 16 K of memory to store a screenful of data. This is just what is done with a memory-mapped VDU. Everything that appears on the screen is held in a special area of RAM known as the *screen memory*. This area of RAM is so organised that each column of every character's allotted space on the screen has its own unique memory location.

For example, the 16 K block of memory reserved for screen use may be located at the top of the memory map (using, say, a Z80-based machine). This would occupy the addresses $C000_H$–$FFFF_H$ inclusive. The first address, $C000_H$ is at top-left and the last address, $FFFF_H$, is at bottom-right. The first byte of memory at $C000_H$ represents the first 8 horizontal pixels across the screen, the second byte represents the next 8 pixels, and so on for the remaining bytes that make up the total of 80 bytes for the first line. It can be seen from this, that to make up a single character, eight rows of eight pixels (8 bytes) will be required. Each of the eight bytes will have locations within memory 80 bytes apart, that is, they are on successive lines, one byte *under* another.

Figure 6.6 illustrates the point by showing the data loaded into memory locations $C000_H$, $C050_H$, $C0A0_H$, $C0F0_H$, $C140_H$, $C190_H$, $C1E0_H$, $C230_H$, in order to place a capital A at the top left hand corner of the screen. Easy isn't it?

Computer interfacing

Regarding the screen in this way, as just another area of memory, makes it easier to appreciate how any shape can be placed where wanted (by loading the relevant data into the required screen locations) and how this shape can be moved about (by moving the data for it about in screen memory).

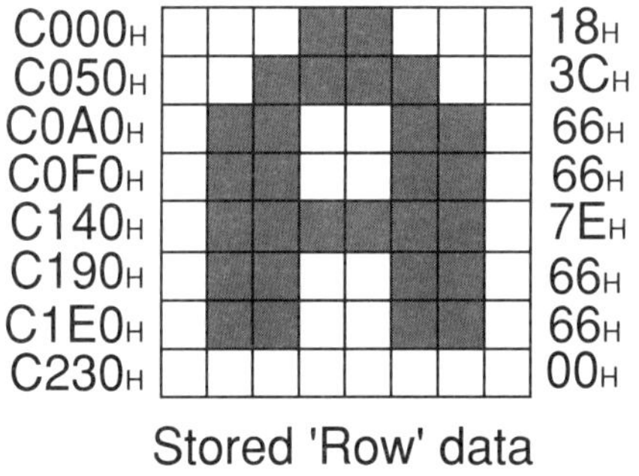

C000H	18H
C050H	3CH
C0A0H	66H
C0F0H	66H
C140H	7EH
C190H	66H
C1E0H	66H
C230H	00H

Stored 'Row' data

Figure 6.6 Capital A: it's dot matrix form, and the HEX column data stored in the screen RAM (addresses $C000_H$, $C050_H$, $C0A0_H$, $C0F0_H$, $C140_H$, $C190_H$, $C1E0_H$ and $C230_H$) as an example of a memory-mapped VDU principle

There is still the matter of producing a steady display on the screen. To do this means that the screen memory must be *scanned* regularly in some way, the rate at which this scanning takes place being synchronised with the rate at which the electron beam scans the screen. If data is changed between one scan and the next then the screen display will change accordingly. As the user types in text from the keyboard, data for this text is entered

120

into screen memory. As the latter is scanned, new characters appear on the screen. Naturally, all of this process happens so quickly that it appears continuous.

The complexities of VDU scanning would occupy more space than is available here. The foregoing applies to specific *dot graphics* displays. If only the standard ASCII characters need to be shown, along with some spare graphics shape characters perhaps, then it is not necessary to use up so much memory to literally *draw* each character in such fine detail. Instead, screen RAM can merely comprise a number of locations which contain the ASCII values of the characters in the required places, and for an 80×25 character screen say, only 2000 actual bytes of screen RAM are needed for this, a great saving on memory usage. The shape of each actual character is stored in a separate *character ROM*, which is addressed by a combination of the ASCII value plus a value of 0 to 7. The character ROM then has eight bytes for each character, each byte representing 8 dots (bits) which makes up one eighth of the character — the eight bytes are organised from top to bottom (eight dot rows). The VDU display circuit scans each text line in screen RAM eight times, to produce eight equivalent TV scan lines, using the ASCII value of each character in turn to find the right letter shape — and the scan line number (one of eight) to find the right row — in the character ROM. The relevant byte (row of dots) is loaded into a shift register and transmitted in serial form to the VDU by a high frequency counter commonly known as a *dot clock*.

It can be seen that a lot of the work is carried out by a dedicated CRTC (cathode ray tube controller) IC, such as the 6845. This IC has 18 registers whose contents are controlled by the ROM-based operating system. But such a display controller is invariably *self running* so all that the user (and the computer come to that) has to worry about is what to put in the screen RAM in the first place!

7 Magnetic recording principles and methods

The basic principle of recording on the surface of a disk, whether hard or floppy, is similar to that of recording on audio tape. A head, consisting of a small coil on a ferro-magnetic armature, is positioned either in contact with the surface of the medium or very close to it. This head, having a dual role, is termed the read/write head. During the *write* operation it carries a current (I_w in Figure 7.1) so as to produce local areas of magnetisation in the surface of the magnetic medium as the latter passes beneath it. These are known as *bit cells*, as each stores one binary digit, or bit, of information. Naturally, with up to eight million such cells on some types of floppy

Computer interfacing

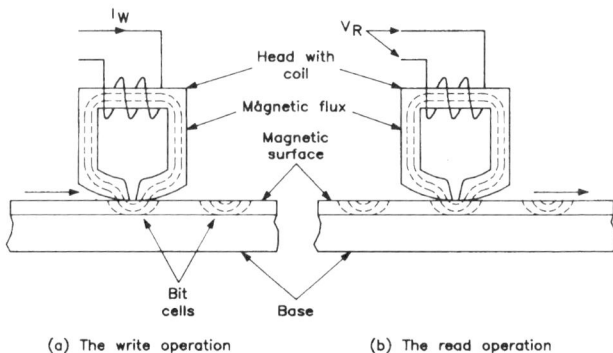

(a) The write operation   (b) The read operation

Figure 7.1   Action of the read/write head in writing or reading data in bit cells

disk, they are rather small in size! During the *read operation*, the head will, by induction, pick up a minute voltage (V_R in Figure 7.1) from one of these bit cells as it passes over it.

Recording codes

There is a number of ways of storing binary digits in the bit cells referred to. These are illustrated in Figure 7.2, to which the following notes refer:

(a)   *return to zero* (RZ): a logic 1 is represented by one magnetic state for part of the cell, the rest being returned to zero magnetisation. A logic 0 is represented by *no magnetisation* throughout the cell. The write head supplies a short current pulse to write a 1,

(b)   *return to saturation* (RS): a logic 1 is represented by one magnetic state for part of the cell, the rest being returned to the opposite magnetic state. A logic 0 is represented by no change from the latter state throughout the cell,

(c)   *bipolar return to zero* (BRZ): a logic 1 is represented by one magnetic state (e.g. a North pole) for part of the cell; a logic 0 is represented by the opposite magnetic state (South pole), also for part of a cell. In both cases, for the rest of the cell the magnetisation is zero,

(d)   *non-return to zero* (NRZ): a logic 1 is represented by one magnetic state, a logic 0 by the other magnetic state. In both cases the whole of the cell is magnetised,

(e)   *non-return to zero 1* or *non-return to zero invert* (NRZI): a logic 1 is represented by a change in the magnetic flux (in either direction) at the beginning of the bit cell, while logic 0 is represented by no change,

(f)   *frequency modulation* (FM): at the beginning of every bit cell there is a flux change; for a logic 1 there is also a flux change at the middle of each cell, while for a logic 0 there is no flux change at the middle of the cell. This is the method commonly used for *single-density* recording,

(g)   *modified frequency modulation* (MFM): the logic 1 and 0 states are defined exactly as for the case of FM. The difference lies in the fact that there is only a flux change at the beginning of a bit cell if both the previous and present bits are logic 0s. This is the method used for *double density* recording,

(h)   *phase encoding* (PE): a logic 1 is represented by a *change in the flux direction*; a logic 0 is then represented by a change of flux in the opposite direction.

Computer interfacing

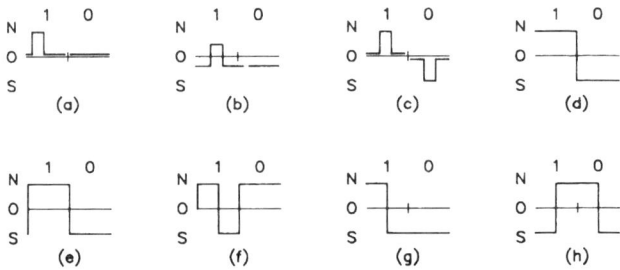

Figure 7.2   Some of the large variety of codes used in disk recording

The reader may be interested (and amazed!) to discover that there are several other codes for disk recording, but this review should suffice to give an idea of the various techniques.

Floppy disks and floppy disk systems

Floppy disks, also known sometimes as *diskettes*, form a convenient mass storage medium for the microcomputer. They are often referred to as *backing store*, meaning that they are additional to the storage provided internally by the RAM area of computer memory. Not only can the user employ floppy disks to store his own programs and data, but much commercial software, including a large amount in the public domain, is available in this medium also. The speed with which this data can be accessed is quite adequate for the majority of applications. The price of disks has reduced substantially during recent years, another factor in their favour.

126

Magnetic recording principles and methods

If there is one problem with floppy disks it is the perennial one of *standards*. The original floppy disks were quite large, about eight inches in diameter. Following this was the size that itself became virtually a standard, certainly among the smaller users, the 5.25 in standard, an increasingly large number of users now opt for the *new standard* of 3.5 in disks. Indeed, some PCs have both sizes of drive unit fitted so that either disk can be accepted. There are also ways in which data can be transferred between the two standards. The capacity of the 3.5 in disk is comparable to that of the larger size, in excess 1 megabyte of total storage on the two sides. The smaller disk is more robust, being protected in a hard plastic sleeve, unlike the thin card sleeve of the larger disk. The price has also fallen to a comparable level.

Irrespective of the size of disk considered, the principle is the same. The *disk* itself is made of a thin, flexible plastic, the surface of which is coated with a magnetic material in which the data is recorded by means of a read/ write head that bears lightly on the surface. This disk, when it is being accessed in the drive unit, rotates within its protective sleeve at a speed of 300 rpm. Cut out of this sleeve are two *windows*. One of these is a radial slot that provides the required access to the magnetic surface for the read/write head; the other is a small index hole which provides a positional reference or datum point. In the case of the 5.25 in disks, these windows are permanently open, hence the need for a paper pocket into which the disk is placed when not in use. In the case of the smaller disks, these windows are protected by spring-loaded metal shutters that open automatically when the disk is placed in the drive. The position of the index hole is identified by an optical device.

Computer interfacing

In the same way that audio and video cassettes can be protected from *recording over* by breaking out a tab, the absence or presence of which a sensor in the machine detects, computer disks can be *write protected* either by a small patch placed over the write-protect notch at the side of the disk, in the case of 5.25 in, or by moving a small plastic slide, that opens or closes a small circular hole, in the case of 3.5 in disks.

Variations in disks, apart from their size, include whether they are single or double-sided and whether single or double density. The data is recorded in the form of concentric tracks, of which there are usually either 40 or 80. The outermost track is generally called *track 0*. The two sides of a 5.25 in and 3.5 in disk may be accessed by independent heads, so avoiding the necessity for turning the disk over in use. By contrast, drives using the less popular 3 in standard are invariably single-sided, so that to access the alternative side, the disk must be physically removed from the drive, *flipped* over and re-inserted.

As already stated, the magnetic recording surface is divided into concentric tracks, up to 80. A further division is made, radially into segments, producing sectors of a typical storage capacity of either 256 or 512 bytes. Each of these sectors is identified by address information stored at their intersections. This sub-division of the surface into small, individual storage areas has no physical reality. It exists only because of the magnetic pattern *imprinted* on the disk surface by a process known as *formatting*. A new disk has to be put through this process before being used for the first time. Thus, when it is then put into service, the magnetic pattern is recognised and

128

the allotted positions of the sectors identified. This arrangement of tracks and sectors is illustrated in Figure 7.3.

A 5.25 in disk, formatted for 80 tracks per side, has a track packing density of 96 tracks per inch (96 TPI). Thus, only the outer area of the disk is actually used for the recording of data, otherwise the sectors near the centre of the disk would become unreasonably small and cramped. The two sides of a disk may be known as sides 0 and 1. Consecutively numbering all tracks on both sides of the disk would give the following arrangement:

● on side 0: tracks 0 (outermost) to 79 (innermost),

● on side 1: tracks 80 (innermost) to 159 (outermost).

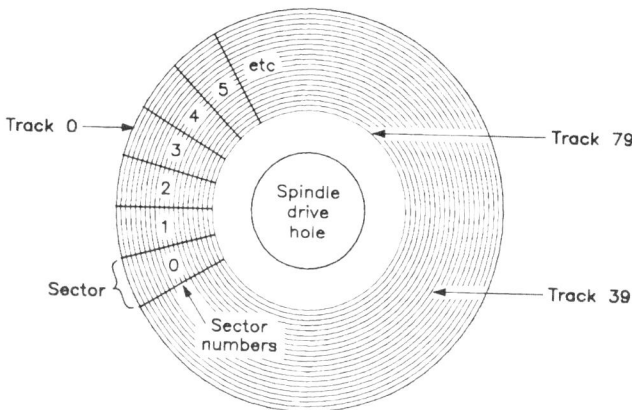

Figure 7.3 Typical arrangement of tracks and sectors on a formatted disk

Computer interfacing

A 5.25 in disk will be formatted to have either 40 or 80 tracks per side with 10 sectors per track. Assuming each sector holds 256 bytes of data so that, with an 80 track, single-density recording, the total capacity is: $256 \times 80 \times 10 \times 2 = 400$ kilobytes. Using double-density recording the capacity becomes 800 kilobytes, naturally.

Files and directories

Programs and data stored on disks are held in blocks whose size, in K-bytes, depends upon the size of the program or data file. Each block has its own unique track/sector start address and a *filename*. The latter is usually restricted to a maximum of eight alpha-numeric characters plus a three character extension. The term *file* is, thus, quite general for any recorded block of data, regardless of its nature or function. To keep track of names and locations, a directory is created on the disk, usually in the outermost tracks. This directory usually has a limited capacity, thus restricting the number of files that can be held on disk, regardless of the capacity of the disk or the size of the individual files.

When a *read* operation is initiated for a specified file, the directory is first searched for a file of that name. If no such file exists in the directory an error message is displayed. If the file does exist, then the location and number of sectors comprising the file are read from the information held in the directory and the transfer begins.

Magnetic recording principles and methods

When a *write* operation takes place, the new file is recorded at the next free sector onwards, and its name and other details are written into the directory. In the event that a file of that name already exists, the original file may be modified to include a .BAK extension to the end of its file name, and in this original form will still exist while the new version is added to the disk.

The operation of deliberately erasing a file, say by issuing the ERASE command, does not actually affect the recorded data at all. It merely erases the entry for that file in the directory, so allowing further files to overwrite the original file space. Only then is the original data lost. The use of a disk editor, after erasing a file but prior to actually overwriting it, can allow the file to be recovered by replacing the directory entry.

The floppy disk drive unit

The drive houses both the mechanical functions and the electronics required to rotate the disk and access the data via the movable read/write head. The head assembly moves radially across the disk surface, actuated by a head positioning worm screw arrangement, endless loop belt or some such that is driven by a stepper motor. In this way the head is accurately positioned above the required track, while the sectors are identified, as the disk rotates, by their positions relative to the index hole mentioned previously. The position of the latter is sensed optically by a photo-cell. The basic form of a floppy disk drive unit is shown in Figure 7.4.

Computer interfacing

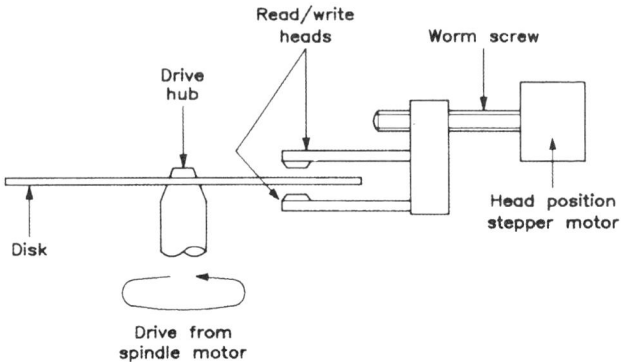

Figure 7.4   Simple mechanical concept of a floppy disk drive
unit

The drive electronics must perform a number of tasks:

(a)   move the head to the required track,

(b)   load the head and set up for either reading or writing,

(c)   generate or recognise various control signals, such as those that identify track 0 and the location of the index hole,

(d)   drive the spindle motor at an accurate rotational speed.

During a data transfer there are three identifiable operations, namely:

(a)   head positioning,

(b)   read/write control,

(c)   the actual data transfer.

132

Magnetic recording principles and methods

As already stated the head is positioned by a stepper motor. This type of motor moves in specific increments of *so many degrees* for each applied pulse. The program that controls this motor must, therefore, generate a specific number of pulses in order to cause the stepper motor to rotate through a particular angle. The positioning mechanism converts this rotation into the required linear movement of the head across the tracks of the disk to the one that has been selected. The head is then *loaded* onto the disk, that is, it is lowered onto its surface. Once loaded, the track number is read to verify position, by comparison with the track register.

Reference to the other two operations is made later.

The disk drive interface

Interfacing a microprocessor to a disk drive unit is performed by means of a dedicated control IC and interface unit. A typical floppy disk controller IC is shown in Figure 7.5.

Software sends to the controller IC the required track and sector address. The IC then sets the *head direction* and *head step* signals to position the read/write head correctly. The pulse generated by the index hole is used to determine the correct angular excitation and, hence, the required sector. When the head is correctly positioned at the required sector/track address, the *head load* signal is *set* and the head is then lowered onto the disk surface. The *read/write* signal determines the direction of data transfer. Data is, of course, written to, or

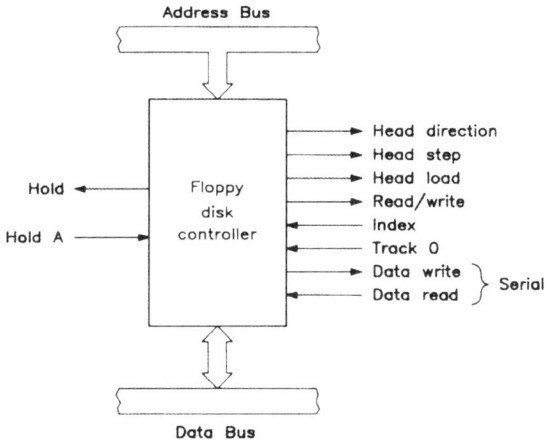

Figure 7.5 Schematic for a floppy disk controller IC

read from, the disk in serial form, bit by bit. Another of the functions of the controller IC is to perform serial-parallel conversion (during the *read* operation) and parallel-serial conversion during the *write* operation. When the disk drive is first switched on, the read/write head sets itself to the track 0 position; at this time, the *track 0* signal is used to reset the IC's track register.

There are essentially two ways of transferring data between the computer memory and disk drive unit. One way is to use normal input/output techniques, the transfers, byte by byte, being handled by a software program. This tends to be slow because of the numbers of instructions that have to be executed during transfers of substantial amounts of data.

The favoured method, that avoids this, is *direct memory access* (DMA). In this method, the transfer of a specified block of data is initiated by software but a hardware device, called a *floppy disk controller* (FDC) IC, takes over. This handles the transfer itself without any further software intervention. Figure 7.6 illustrates the basic idea behind the DMA method, showing how the CPU is bypassed during transfers.

The FDC provides the necessary interface between the CPU and the disk drive, converting the software commands issued by the *disk operating system* (DOS) into the electrical signals needed to control the drive. Special-purpose ICs, such as Intel's 8272 exist for this purpose. It is possible to have what is known as an *intelligent floppy disk interface*, which has its own on-board CPU, such as a Z80.

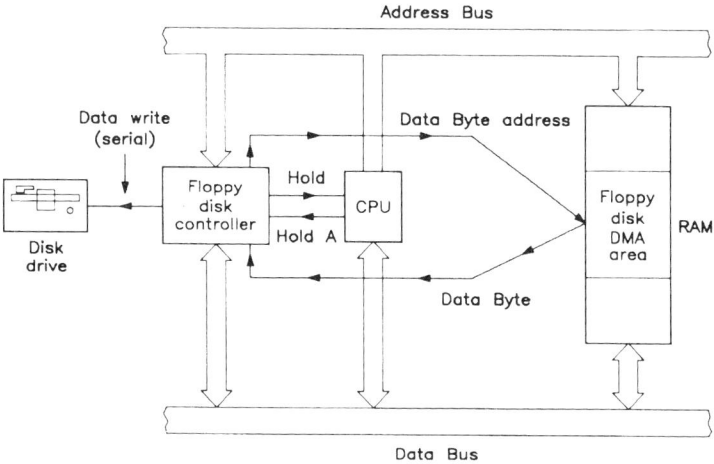

Figure 7.6   A Direct Memory Access (DMA) system in use, showing how the CPU is effectively bypassed

Computer interfacing

Basic functions of a floppy disk controller

These are as follows:

(a)   *select drive*: one controller may look after a maximum of four drives, with the ability to select any one on receipt of the appropriate command,

(b)   *seek track*: a sequence of step pulses moves the read/write head to the required track, in a particular direction. If moving outwards, the track 0 signal is used to terminate the movement when this track is reached,

(c)   *read/write*: the controller *loads* the head onto the disk surface, allows a *settling time* before the transfer commences and generates a signal to select either the read or write mode,

(d)   *data separation*: as recorded, the data to/from disk is a serial bit stream of mixed data and clock pulses. The actual data must be separated from the clock pulses then converted to parallel form,

(e)   *error detection*: in any serial data path, such as that between memory and disk drive, there is always the possibility of data being *corrupted*, that is, individual bits inverting their binary values. Received data is then incorrect but unless this fact is detected in some way, can be accepted as being valid. Writing to disk is normally error checked by reading during the next revolution of the disk. After, say, 10 such attempts have failed to produce valid data, an error message should be displayed that may even identify the faulty sector.

Errors may be caused by electrical noise that, if temporary, will eventually allow a successful transfer; this is

136

known as a *soft* error. Hard errors, on the other hand, are usually permanent and may be caused by physical contamination of the disk surface. Error checking is usually performed by a sophisticated checksum method known as the *cyclic redundancy check* (CRC).

Operation of the floppy disk controller

Taking the *write* operation as an example (data being *written* from computer memory to disk), the role of the software is to:

(a)   supply the floppy disk controller IC with the start address (in an area of the computer RAM memory known as the DMA area) of the data block to be transferred,

(b)   supply the corresponding start address on the disk where the data is to be stored; this takes the form of track/sector information. Also required is the number of sectors needed for the storage.

On receipt of this information, the FDC IC takes over. Its first function is to locate the specified track/sector address. Upon doing so, it generates and sends a HOLD signal to the CPU. The latter completes its current instruction and responds with a HOLDA (*hold acknowledge*) signal. The FDC takes over direct control of both the data and address buses and carries out the data transfer. Unlike the normal input/output transfers between CPU and peripherals, which use the accumulator (A-register) as a *transit area* for the input/output ports, the CPU is completely bypassed in DMA transfers. This speeds up the transfer rate enormously.

Hard disk systems

Hard disks, also known as *Winchesters*, are of rigid construction, hence the name. The disks are usually made from aluminium, coated with a magnetic material such as ferric oxide or chromium oxide. The special aerodynamically shaped head is known as a *flying head*, and fly heights of 0.5–3 microns are usual (a micron is a millionth part of a metre). The heads can be designed to take off from, and land on the disk surface, this feature being characteristic of Winchester drives. The disk surface is lubricated to minimise the risk of damage. The complete system is hermetically sealed to provide a dust-free environment.

Unlike the floppy disk, which is only rotated during access times, the hard disk assembly rotates continuously, at a speed of 3600 rpm. Because of their bulk and consequent inertia, it can take quite a few seconds for hard disks to reach their full operating speed.

Hard disk systems may use either fixed or moving heads. Fixed head disks have one read/write head per track, an extravagance that is compensated for by the much reduced access time and the saving made by not requiring a positional motor drive. A moving head assembly uses one read/write head per disk surface, thus requiring a positional motor for accessing the required track on a surface, in exactly the same manner as for floppy disk systems.

Advantages of hard disks over floppies are:

(a)   a much greater storage capacity. A 120 megabyte system is now considered quite modest while, for those who can afford them, hard disk units offering several gigabytes of storage are now available,

(b)   a faster access time, by a factor of at least 10:1, compared with floppy disks,

(c)   system software is automatically loaded when the system is switched on, giving immediate access to application software such as word processors, databases, etc.

Naturally, there are different sizes of hard disk, according to capacity, but the user may be less aware of this due to the fact that the hard disk unit is housed within the main computer casing. Large capacity Winchesters are of *multi-platter* construction. The 22 megabyte disk assembly shown in Figure 7.7 has four double-sided platters, thus requiring eight read/write heads altogether.

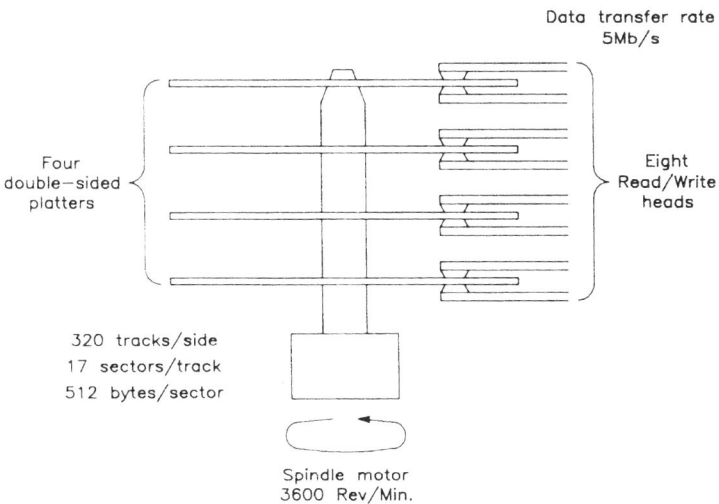

Figure 7.7   The layout of heads and platter for one form of 22 megabyte hard disk drive

Computer interfacing

Each platter has 320 tracks per side with 17 sectors per track, each sector having a capacity of 512 bytes. Using this data, the calculation for the total storage capacity is: total capacity in kilobytes = $8 \times 320 \times 17 \times 0.5$ = 21760 kilobytes, (512 bytes = 0.5 kilobytes), which equals 22 megabytes approximately.

There is a standard connector interface for Winchesters, the details being given in Figure 7.8.

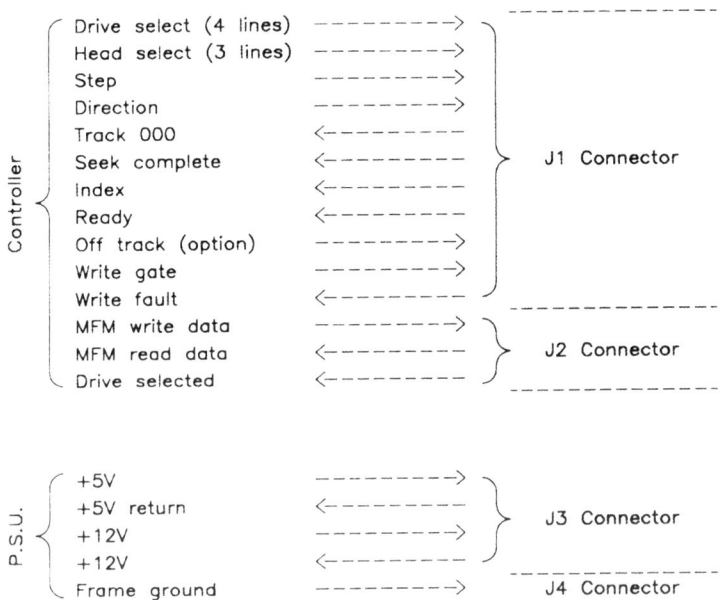

Figure 7.8   Standard interface connections for a Winchester (hard disk) drive system

8 Motor drives

Many applications of microcomputers result in an output to drive a motor. Without stopping to think about this fact, it may not seem immediately obvious. After all, word processors and databases seem totally divorced from the world of industrial machine control. Yet motors of one sort or another are frequently essential in both.

To take the use of a word processor, as a familiar example. No less than *four* separate motors are involved in the process! The software to run the program is invariably contained on disk (don't write in if yours is in ROM!), whether it is a floppy or a hard disk. To access the application program itself, as well as the files for the work being done, requires two motors, as follows:

Computer interfacing

(a)   one to spin the disk itself at 360 rpm,

(b)   one to drive the read/write head radially across the disk surface to find the required tracks and sectors where the data is stored.

As a result of the word processor being used to produce a document, whether a short letter or a block-busting novel, a printer will be used to turn the *silicon copy* into something rather more permanent and substantial. This also requires two motors:

(c)   one to rotate the platen, to feed the paper,

(d)   one to drive the print head across the paper as the characters are struck onto it. This assumes a dot-matrix printer, which doesn't need a motor to rotate its print head.

In fact, applications of this nature may actually make more use of motors than rather more *physical* tasks. In some of the latter cases no motors are needed at all. A temperature control system, for example, unless it uses an output device such as a printer for a hard copy record of temperature/time variations, is merely converting between analogue and digital quantities (and vice-versa), purely electronically.

Motors are, traditionally, analogue devices, or rather they were until relatively recently. Because the computer can only input and output digital data, there would seem to be a need for conversion when a conventional motor is to be driven by a computer. As it happens, there are several choices, as follows:

(a)   use a conventional motor and a digital-to-analogue converter,

142

(b)   use a conventional motor and drive it *digitally*,

(c)   use a motor that takes a digital input directly.

Figure 8.1 shows the first case. A digital quantity, representing motor speed, is delivered to a digital-to-analogue converter (DAC). The analogue output drives current through the field winding and, hence, determines the motor speed. Naturally, the binary output will be generated by a software program. This, in its turn, may have received input data — perhaps from a set of switches at an input port — that demands a certain motor speed.

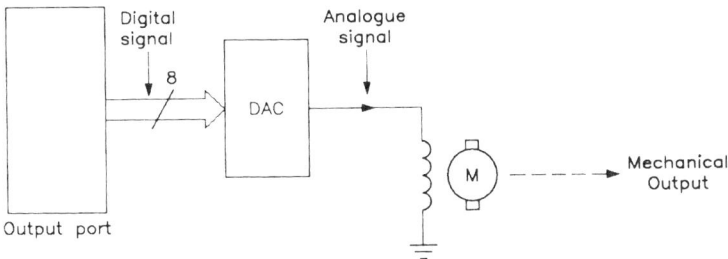

Figure 8.1   Using a DAC-driven analogue voltage to drive an analogue motor

Figure 8.2 shows the same motor being used, but instead of the supply being an analogue quantity, it is a square-wave of variable mark/space ratio. The speed depends upon the ratio of on-time/off-time. It is easiest if the off-time is a constant value and the on-time only is varied.

143

Computer interfacing

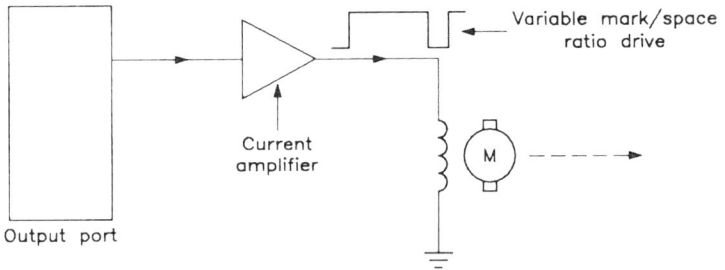

Figure 8.2   So called digital control of an analogue motor

Figure 8.3 shows two possible ways of using inputs to control motor speed. In the first case, (i), a potentiometer is used as the speed controller; its wiper voltage is converted to a digital value which the program reads at the input port. This is used to set the length of the time waste loop and, hence, mark/space ratio and motor speed.

In the second method, (ii), the speed is selected by setting combinations of switches. For example, if three switches are used, they can be set in $2^3 = 8$ different combinations. This means that the choice of speeds is limited to this number. Whether this is adequate depends upon the application. In case it might be thought that using switches in this way to control motor speed is somewhat cumbersome, it might be worth bearing in mind that such switches might not be *front panel* switches at all but *embedded* in the process (whatever that is). For example, the switches might be operated in a certain sequence that depends upon the nature of the process. As each stage is completed (signified by the closing of a switch), a new motor speed (higher or lower) is selected automatically for the next stage.

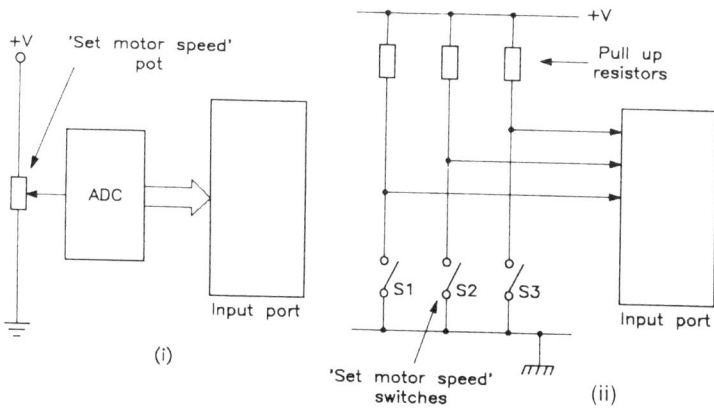

Figure 8.3 (i) Continuous speed control system (ii) Stepped speed control system with binary switches

Using the method of Figure 8.3 (i) gives almost *stepless* speed control. If an 8-bit converter is used, this gives 2^8 = 256 different values (speeds) between zero and maximum. Thus, rotation of the potentiometer gives smooth speed control.

This brings us to the third option, a motor that takes a digital input directly. An example of a motor of this type is the *stepper* motor. In practice, it is this type of motor that is generally employed in the applications mentioned at the start of this chapter, namely in disk drives and printers. Bearing in mind what was said about these applications, it is possible to identify two different ways in which motors are used, in fact two different types of system:

(a)   a speed control system, such as when controlling the rotational speed of the disk at exactly 360 rpm,

145

Computer interfacing

(b) a positional control system, as when setting the read/write head correctly over the track (disk drive) or when moving the platen or print head just the right amount, as in printers.

The stepper motor is able to handle both of these tasks with great accuracy and without any really expensive electronics either. This is, of course, the prime reason for its use, apart from the obvious fact of not needing any signal conversion. Before looking at the stepper motor in principle and application, it is worth seeing the problems that exist in the two types of system mentioned and how they are handled in an analogue system.

Figure 8.4 shows an analogue speed control system. The motor is driven from the output of a power amplifier. The input to this amplifier is the difference between two voltages (V_P-V_G) and is known as the *error* voltage. One of these voltages, namely V_P, represents *speed demanded* and is sent in from a potentiometer. The second voltage, V_G, which only exists when the motor is running, is the output from a small generator (known as a tachometer or *tacho*), driven by the motor.

Assume the system is at rest and power is then applied. With the potentiometer wiper anywhere other than at the bottom of the track, there is an input V_P, to the amplifier, an output from the amplifier and plenty of torque to accelerate the motor. As the latter picks up speed, the tacho output V_G increases. As the latter voltage is in anti-phase to the potentiometer voltage, the error voltage decreases. A balance is reached when the difference between the input and tacho voltages is just enough to maintain a given motor speed. Any change in the input

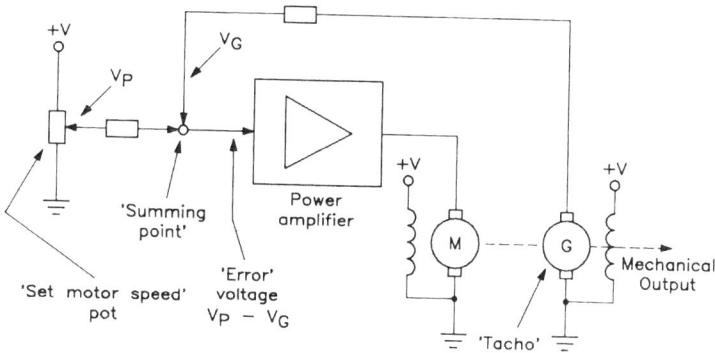

Figure 8.4 An analogue speed control servo system

demand changes the error voltage, up or down until, at a different speed, balance is reached again. Notice that the system is self-balancing due to the negative feedback provided by the tacho. Such a system has the general title of *speed control servo system*.

Figure 8.5 shows another servo system, this time for positional control. In this diagram there is a potentiometer input as before. The difference is that, in this case, it is not demanding a particular speed of the output shaft, but a particular position. The way it does this is also by using negative feedback, from a *second* potentiometer which has its voltage applied in the opposite polarity. In fact, both input and feedback potentiometers are usually centre-tapped to allow control in both directions on either side of centre. Ignoring the tacho in the diagram for the moment, the system works as follows.

The input to the amplifier is the difference between the two wiper voltages, V_1 and V_2. When both shafts are

147

Computer interfacing

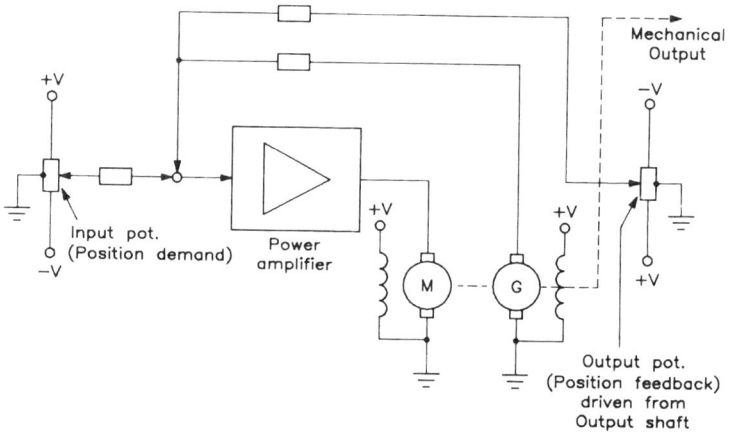

Figure 8.5 An analogue position control system

aligned the two wiper voltages are equal and opposite and cancel out. There is no input to the amplifier and no output to drive the motor. The system is at rest. Now rotate the input shaft; immediately there is an error voltage, V_1–V_2, which is amplified to drive the motor. As the latter runs it drives the output shaft in such a direction that a position will be reached when, once again, input and output shafts are aligned and there is no error voltage. The system comes to rest.

But, what is the tacho doing? Its role is not the same as in the case of the speed control system, where it had to provide a feedback voltage to maintain constant speed. In this case its function is to provide a *damping* term, a rather more obscure function. Here's how it works.

Imagine that the tacho is omitted. It would seem that the system would still work. However, what is ignored is the effect of system inertia, or momentum if you like. When

148

a motor is accelerating, as it does when it is racing towards its new position, it acquires momentum because of its mass. Just because it has reached its new position and no longer has a signal from the amplifier to drive it, doesn't mean that it will suddenly stop. Far from it; its momentum will carry it past the correct position. As a result of this overshoot, a feedback voltage will be derived from the wiper of the output potentiometer that will apply a torque to the motor *in the opposite direction*. The motor stops (too late!), reverses, and heads back the way it came, once more gaining momentum. This time it overshoots from the opposite direction, probably not quite so much. The process of continually overshooting repeats, again and again. In fact the system may become oscillatory.

What is needed is a means of decelerating the motor as it approaches its correct position, in fact a form of brake. The output from the tacho provides this, as it opposes the driving torque. As the motor gets less drive from the amplifier (approaching final position), the tacho voltage has more effect and brakes the motor. With the correct amount of feedback from the tacho, the system does a mild overshoot, returns and stops dead. Too much damping reduces the responsiveness; it becomes sluggish. Too little damping and the system tends to oscillate.

Closed and open loop systems

The types of system described above are known as closed loop systems, because the feedback path *closes* a loop.

Computer interfacing

They can give great precision of control, speed or position, but are complex, require expensive components and can be tricky to set up.

By contrast, Figure 8.6 shows an *open loop* speed control system. There is no feedback, so no self-balancing action. Setting the input wiper sets a certain speed. There is nothing that automatically governs this speed. Variations in load, for example, cause the speed to vary. It is cheaper than an open loop system but not very precise.

Where does the stepper motor come in? Is it an open loop or a closed loop device? The answer is, it is an open loop device. Yet, because it works in a totally different way from an analogue motor, it is capable of excellent precision and doesn't require complex electronics to drive it. Nor is the motor itself inherently expensive.

Its secret can be understood from the fact that it is *pulsed* in order to produce specific angles of rotation. For example, if one pulse caused the shaft to rotate *10 degrees,*

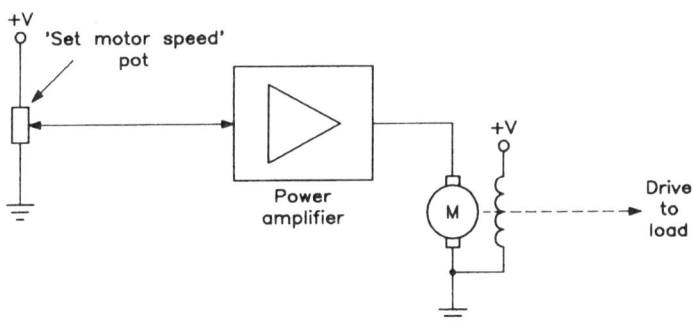

Figure 8.6 An open loop speed control system

then in order to rotate it through exactly 40 degrees would involve nothing more than giving it four pulses. No elaborate feedback arrangements are needed.

Naturally, there are some complications and practical motors need a little more than a single pulsed input. But stepper motor controllers and drivers are available in IC form, requiring only a few external components. So now for some basic principles.

Stepper motors

A stepper motor consists of a stator, rotor and windings. It does not rotate when power is applied but moves through a precise angle in response to a single pulse. Therefore the total angular displacement is given by: total angular displacement = step angle × number of pulses.

Step angles vary in the range 0.45° to 90°, the most common value being 1.8° per step (giving 200 steps per revolution). A typical positional error is ±5%.

Stepper motors can be used for linear rather than angular positioning by using one to drive a leadscrew and nut. The relation between the linear displacement and the number of step pulses is such that: linear displacement equals the number of steps divided by the total number of steps per revolution multiplied by the pitch of the lead screw. For example: if the leadscrew has 5 threads/inch, its pitch is $\frac{1}{5}$ in or 0.2 in. If the step angle is 1.8°, each step pulse produces a linear displacement of 1 ÷ 200 x 0.2 = 0.001 in.

Computer interfacing

This makes the stepper motor suitable for industrial applications in such things as automatic lathes, milling and drilling machines, X-Y co-ordinate tables and other positioning mechanisms. This is in addition to the computer applications already discussed.

Advantages of stepper motor systems

The major advantage of stepper motor systems is that control is *open loop*, so avoiding the problems of closed loop control. All that is needed is a counter (software or hardware) that counts the number of pulses from the start. The speed of rotation is controlled by the frequency of the pulses, and the acceleration is determined by the rate of change of frequency — see Figure 8.7.

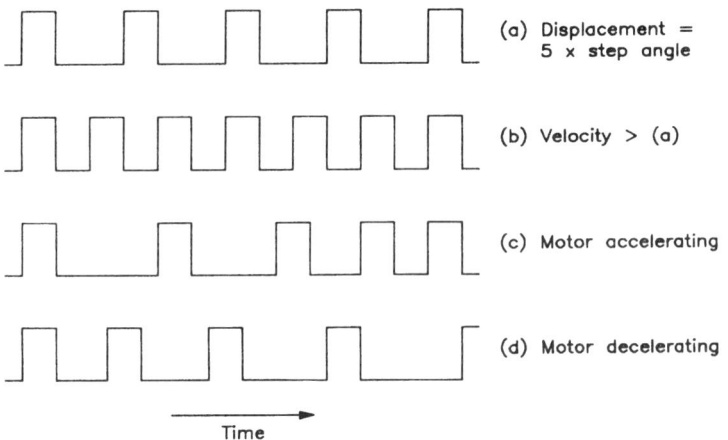

(a) Displacement = 5 x step angle

(b) Velocity > (a)

(c) Motor accelerating

(d) Motor decelerating

Time

Figure 8.7   Pulses control position, speed and acceleration

Stepper motor principles

There are three common types:

(a)   variable reluctance motor,

(b)   permanent magnet motor,

(c)   hybrid motor.

The variable reluctance motor

This type (Figure 8.8) has a soft iron core with projecting teeth at precise angular intervals. The stator has
inward projecting teeth but there are more of them so
that, when two rotor teeth align with two stator teeth all
the rest are misaligned (by an angle of 15° in the example shown). The stator teeth have coils wound on them,
which can magnetise them if d.c. is passed through them.
In the example, coils A are magnetised. If instead coils D

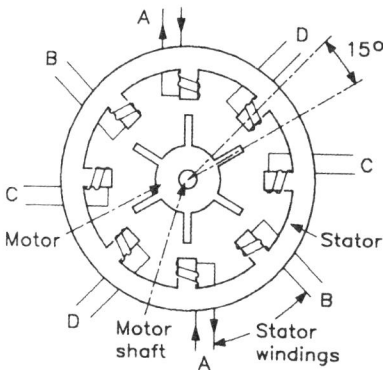

Figure 8.8   Basic four-phase variable reluctance stepper motor

are magnetised, the rotor moves anti-clockwise (ACW) 15°. If coils B are magnetised, the rotor moves 15°. The rotor can be made to rotate one way or the other by sequencing the supply to the coils.

Variable reluctance motors can be used at high speed but with light loads. There is little residual magnetism to hold the rotor in position when the supply is switched off. The inertia of the rotor causes overshoot and oscillation.

The permanent magnet motor

The rotor (Figure 8.9) is a radially magnetised permanent magnet that rotates to align itself with the field produced by the electromagnets. Using the switches A, B, C and D gives four possible field directions and, hence, four positions (90° steps) for the rotor. Half steps can be obtained if the switches are three-position with the centre position *open*.

In Table 8.1 a *1* indicates a phase energised; a *0* is a de-energised phase. This table shows the switching sequence for half-step and full-step operation of a four-phase stepper motor.

The hybrid stepper motor

In this type the rotor has two sets of teeth at 180° to each other and is axially magnetised, one set of teeth being North poles, the other set being South poles. Stator windings can also be either N or S, depending upon the directions of the currents in the coils. Between the rotor and stator there will be, at some places forces of

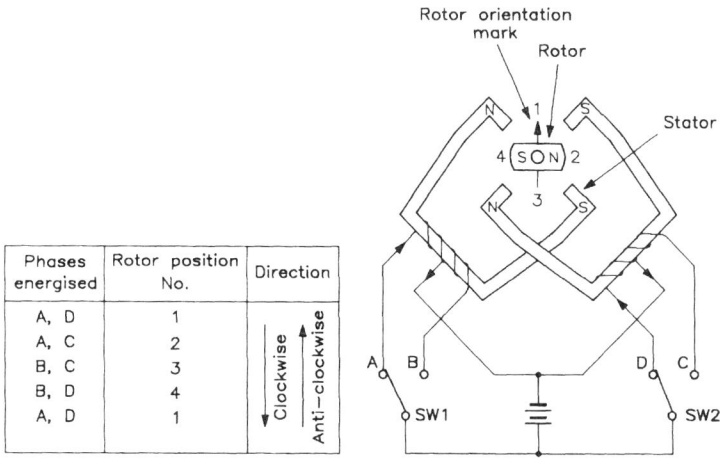

Phases energised	Rotor position No.	Direction
A, D	1	
A, C	2	Clockwise / Anti–clockwise
B, C	3	
B, D	4	
A, D	1	

Figure 8.9 The four-phase, two-pole, permanent magnet stepper motor, showing phase switching to obtain CW or ACW rotation

attraction, and at other places forces of repulsion. These forces cause rotation from one step to the next, controlled by the sequencing of the four phases.

Stepper motor drives

A stepper motor moves from one step to the next by switching a d.c. supply from one set of stator windings to the next. The simplest, and cheapest, way of driving stator windings is with a single drive transistor per winding.

Figure 8.10 shows unipolar and bipolar circuits. The latter type produce 20–40% more power at low speeds than unipolar types but need two power supplies.

155

Computer interfacing

Step	Full step phase			
	A	B	C	D
1	1	0	0	1
2	1	0	1	0
3	0	1	1	0
4	0	1	0	1
1	1	0	0	1

Step	Half step phase			
	A	B	C	D
1	1	0	0	1
2	1	0	0	0
3	1	0	1	0
4	0	0	1	0
5	0	1	1	0
6	0	1	0	0
7	0	1	0	1
8	0	0	0	1
1	1	0	0	1

Table 8.1   Switching sequence of 4-phase stepper motor

Figure 8.11 shows a stepper motor interface; this is a fairly simple type and, in practice, they can be much more complex. The motor windings are numbered 1, 2, 3 and 4.

There are three sequences:

● low power: 1, 2, 3, 4, 1 etc. (only one winding energised),

● normal: 1 and 2; 2 and 3; 3 and 4; 4 and 1, etc. (always two windings energised),

● half-step: 1 and 2; 2; 2 and 3; 3; 3 and 4; 4; 4 and 1; 1, etc. (always half-angle steps between each step).

The motor windings require a current source as their resistance is low, e.g. 0.2 Ω. They also have high inductance values so that protection is needed for the switching transistors.

The timing for mode 2 (normal) is shown in Figure 8.12.

The truth table of Figure 8.12 shows that a half-byte of two 0s and two 1s circulates continuously. This gives a

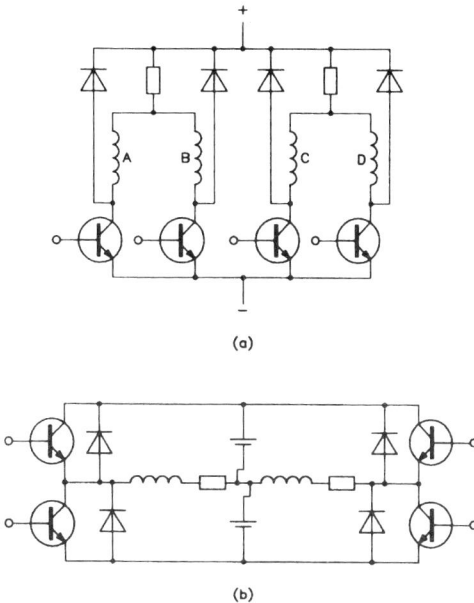

(a)

(b)

Figure 8.10 Drive circuits for (a) unipolar and (b) bipolar stepper motors

Computer interfacing

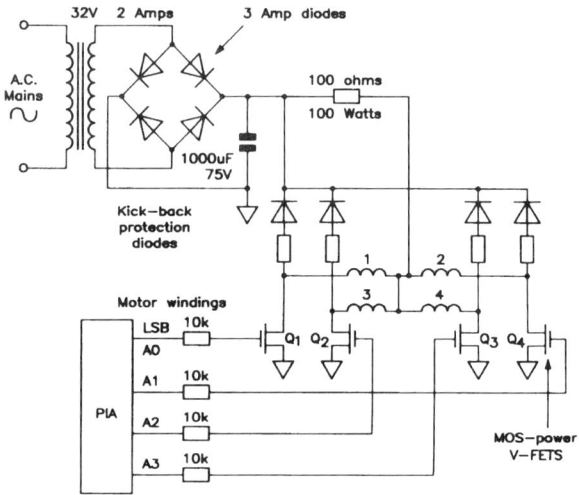

Figure 8.11 A stepper motor interface

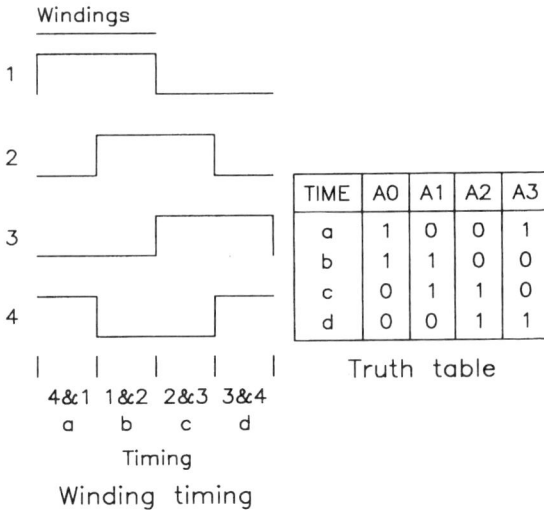

TIME	A0	A1	A2	A3
a	1	0	0	1
b	1	1	0	0
c	0	1	1	0
d	0	0	1	1

Truth table

Winding timing

Figure 8.12 Winding timing diagram and truth table for interface of Figure 8.11

158

clue as to how the waveforms to drive the motor can be obtained. If a 4-stage register is constructed and two successive stages are *set* (Q = 1) while the other two stages are cleared, then clock pulses will circulate the required pattern around the register. Naturally, it is possible to produce the same result by writing a machine code routine that circulates two 1s in this way, outputting them to the PIA between shifts. The number of shifts determine the angle turned through; the frequency of the shifting determines motor speed.

Available stepper motor ICs

The SAA1027 stepper motor driver

This IC is designed to drive 4-phase unipolar stepper motors and operation is simplicity itself.

The logic level on pin-3 determines the direction of rotation. For clockwise rotation it must be low; for anti-clockwise running it should be high. Therefore, under program control the motor is easily reversed merely by writing a different logic level to the PIO port line that feeds this pin A pulse input on pin-15 causes a step whenever the voltage transition is from logic 0 to logic 1.

The L297 and L298 stepper motor controller and driver

The L297 controller works with either 2-phase bipolar or 4-phase unipolar stepper motors. It pairs well with

Computer interfacing

the L298 driver for providing the drive capability to the stepper motor. The application circuit for these two ICs driving a 2-phase stepper motor is shown in Figure 8.13

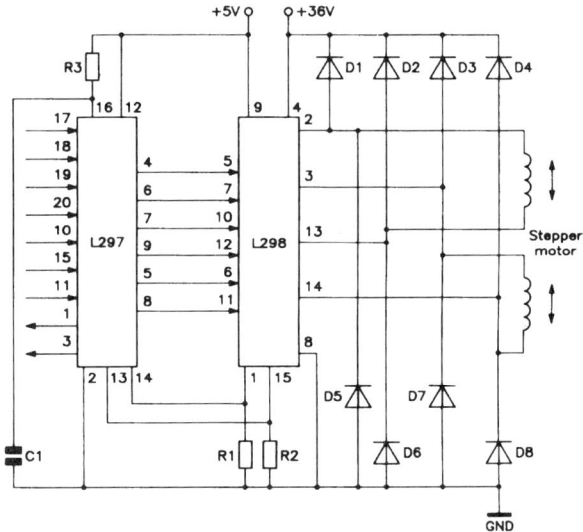

Figure 8.13   Application circuit for the L297 and L298 stepper motor controller and driver ICs

9 Printers

The printer is an example of an *output* device, which the computer drives to provide the user with *hard copy* from his program. A familiar application of this type is word processing. There is little point in writing and manipulating text on-screen if a permanent copy of this cannot afterwards be obtained. Word processing uses may include anything from a short letter, to a block-busting novel. Other software packages that require a printer at the final stage include spreadsheets (for accounting), databases (for record keeping), desk-top publishing or DTP for short (creating a mix of text and graphics for handbills, newsletters, magazines or books for example) and various drawing programs for generating original artwork.

Computer interfacing

To some extent the printer that one buys needs to be matched to the required end product. However, cost influences the choice as well. The price range is wide. It is possible to buy a printer for a little over £100 (the *show bargain* price for some dot matrix models), but more sophisticated models, especially laser printers, can cost anything from £400 to over £2000.

This final chapter takes a look at the various types of printer available and perhaps lifts the lid on a few of their mysteries. There is a certain satisfaction to be gained from listening to the *buzz* of a working printer, while your masterpiece gradually emerges. This is especially true when using the graphics mode, but also when printing text of different styles. There are also great frustrations when things don't go so well; when the printer sits in obstinate silence, when it insists on underlining when it shouldn't or puts in totally uncalled for line or form feeds!

Printers may be placed into three broad classes, termed *character printers*, *line printers* and *page printers*. Most printers fall into the first category; this includes the familiar *dot-matrix* printers and the *daisy-wheel* types. Both of the latter work by printing character by character. The laser printer is an example of a page printer, since it prints a whole page at a time.

If you were to conduct a simple survey of printer types, by reading the advertisement pages of computer magazines, supplemented by a dip into the occasional text book, something like the following list would probably emerge:

- dot-matrix printers, 9-pin or 24-pin,

- daisy-wheel printers,

- laser printers,

- inkjet printers,

- thermal printers.

You might also conclude from such a survey that most printers print in glorious monochrome only (black and white!), but colour printing is also possible, at a price if you really want quality.

The *family tree* of Figure 9.1 shows some attempt to classify the most common types of printer. From this it can be seen that, apart from the classifications of character, line and page made earlier, two other broad classes exist. These cover *impact* and *non-impact* types. These terms are unlikely to confuse anyone.

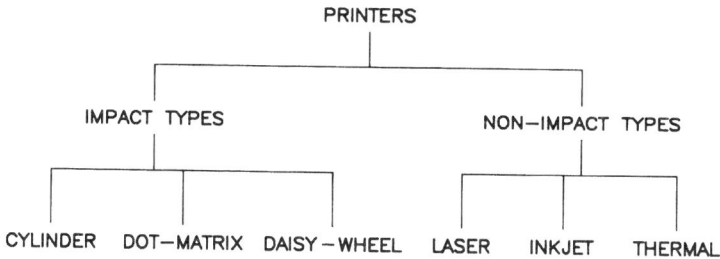

Figure 9.1   Family tree of available printer types

Under the heading of *impact* we find three printers listed; cylinder, dot-matrix and daisy-wheel.

Under the heading of *non-impact* are the three other types; laser, inkjet and thermal.

Computer interfacing

From this selection, the one printer that most current small computer users would like to be using is probably the laser printer, because of its quiet operation and excellent quality of output. Realistically, such users are almost certainly chugging along with a 9-pin dot-matrix type, putting up with the less than perfect type and rather ragged graphics, not forgetting the high pitched buzzing sound that some find very irritating. Certainly, at the present time this type of printer dominates the market. No doubt a changing price structure will cause a swing away from this type at some time.

Cylinder head printers

Our survey of printing methods will start with this one. It is the method that was used by the now obsolete *teletypewriter*, commonly known simply as a *teletype*, such as the ASR-33. Having said that such machines are now outdated, it should be accepted that it is quite likely that a number of them are still going strong, in the hands of dedicated amateurs.

The mechanical arrangement is explained by Figure 9.2. The cylinder of the title has four rows of characters cast into its surface. These are usually limited to capital letters, numerals and punctuation marks. The cylinder is on a vertical pivot and is driven mechanically up and down to select a row and rotated for the required character in that row. Upon selection being completed, a hammer strikes the cylinder, forcing it onto the inked ribbon, thus printing onto the paper. Such machines were

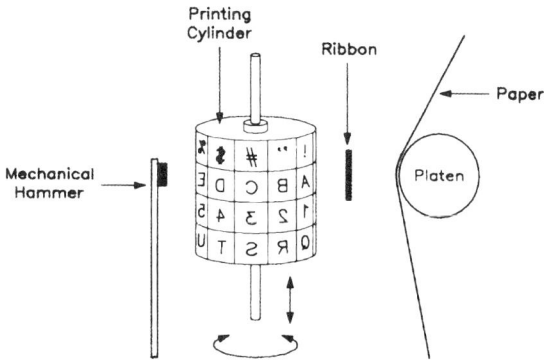

Figure 9.2   The printing mechanism of a cylinder head printer

slow and noisy. The paper normally came in unperfo-
rated, continuous rolls of about 8.5 in width, rather less
convenient perhaps, than the perforated fan-fold paper
in common use today.

Dot-matrix printers

A matrix in this sense is a two-dimensional array of dots,
capable of representing any of a very wide range of characters, alpha-numeric or graphical. The image is
produced by striking the paper through an inked ribbon,
much like that of a typewriter. But here the resemblance
ends. Whilst the typewriter forms each character with a
single stroke, the dot-matrix printer builds up the characters column by column as its print-head passes over
the paper. This print-head incorporates a vertical bank
of fine pins, which can be *fired* electromagnetically at

165

the ribbon. They then return under the control of a spring. In a 9-pin printer there are nine such pins. The matrix for a printer such as the Epson RX80 is nine (vertical) by five (horizontal) dots. To allow for underlining, only the upper area of 7 x 5 dots is used for the character itself. This leaves the 8th row available as a horizontal space, with the 9th row for the underline itself.

Photos 9.l(a) and 9.l(b) show the print-head in close-up. Photo 9.l(a) shows the underside of the print-head, a view not often seen; the pins referred to can be clearly seen. Photo 9.l(b) shows the *working end* of the pins. It may be noticed that several years of regular use have resulted in the ends of some of the pins being somewhat hammered out.

The nature of the character printed depends upon which pins in the 9-pin column are fired at each instant. This probably needs little elaboration as most people regularly see the results of dot-matrix character formation, whether printed on paper or as text on a VDU screen. However, Figure 9.3 shows a typical dot-matrix character in the process of being formed. The character area of 35 pins allows a wide range of typefaces to be created. Examples found on most printers are: pica (normal), elite, italic, enlarged, condensed, double-strike, bold, superscript and subscript. An NLQ (near letter quality) mode is also included nowadays. Some of these modes can be combined e.g. *bold italic* or *condensed enlarged*, thus effectively giving further type faces.

While the 9-pin printer is popular and reasonably cheap, except in its NLQ mode the quality of output leaves a lot to be desired. Apart from using a totally different type

166

Photo 9.1   (a) Underside view of Epson 9-pin printhead showing
the pins themselves (b) the working end of the Epson print-head.
The nine pins are clearly visible, showing some signs of wear

of printer altogether, another possibility is to use a 24-pin dot-matrix printer. These are somewhat dearer, though there are some very reasonably priced models about now. The extra quality obtained by having a larger number of pins is quite dramatic.

It is possible to perform colour printing with this type of printer. A ribbon with three subtractive primary colours; cyan, magenta and yellow, is able to reproduce the three additive primary colours of red, green and blue by over-printing in the appropriate combinations. Often the ribbon is provided with black as well, for normal use. It has to be said that dot-matrix colour printers are quite basic in their capabilities and other technologies have to be employed for any great degree of sophistication.

Dot matrix printers use stepper motors for driving the print-head transversely across the platen, as well as for precise rotations of the platen. This was mentioned in the last chapter, in connection with the uses of stepper motors. Obviously, control of stepper motor position has to be very precise in order that each of the columns that form each character, as well as the spaces between, are printed with exact regularity. Emphasised or double-strike printing is obtained by making one printing pass and then making a second slightly offset from it, so thickening the printed character by printing a second set of dots within the first set. Attempts to obtain a superior appearance, known as NLQ as stated already, rely upon a second pass also, with the dots of the second pass reducing the *stepped* appearance made by the dots in the first pass. The result can be very effective, though still not considered good enough for the best business correspondence.

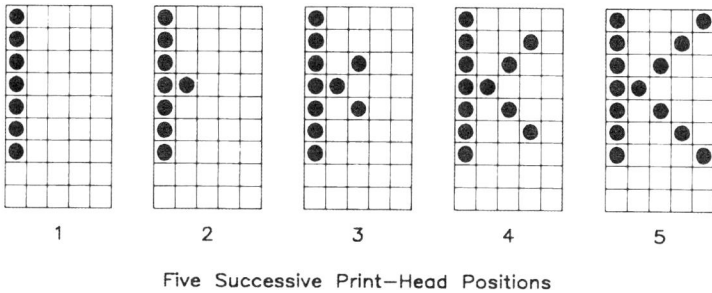

Five Successive Print—Head Positions

Figure 9.3   The dot-matrix character K being formed

In order to carry out any printing operation at all the printer has to be sent a code sequence. A set of character codes defines each character: upper and lower case letters, numerals and punctuation marks; this is the well known ASCII code that we have met before. A set of control codes tells the printer what typeface to print in, when to do a line feed, carriage return, form feed, what spacing to use, where the margins should be and so on. Any variable that you can think of when it comes to printing, is in some way covered by a printer code. When dealing with dot-matrix printers, the standard set of codes are those used by Epson. Most printers conform to these codes. After all, a code is nothing more than a number that defines an operation, so it makes sense that all printers should use the set.

Looking through a printer manual reveals that many of the control codes are of the form ESC x, where x is some alpha-numeric or other character defining the particular function. For example, ESC G turns on double-strike printing and ESC H turns it off again. So, if we want to print a short section of text in double-strike mode, we send the following code sequence to the printer:

27 71 *section of text to be printed* 27 72

The code 27 is ASCII for ESC; 71 and 72 are the ASCII codes for G and H, respectively. Seen in this light, printer control codes have no real mysteries after all! All that is necessary to do, is to insert the required printer control codes, at those points in the supply of data to the printer where certain actions are to be taken. Using a word processor makes this easy; as all such codes are already built in, it is only necessary to make the appropriate keystrokes. For example, using the word processor Protext on an Amstrad PCW 8512, in order to underline a section of text, it is necessary to hit the ALT key followed by the X key (in order to gain access to the printer control codes), then type a *u* (which appears in inverse video) immediately before the text to be underlined. The underlining is turned off by repeating the same sequence.

In this example, typing the first *u* inserts the printer control code for *underlining*; typing the second *u* inserts the printer control code for turning the underlining off. Although such codes appear on the editing screen, they are not printed. They merely control the printing action.

Although many printers are said to be *Epson-compatible*, there are sometimes minor differences. This is also true within the Epson range itself. Some of the codes used by the older MX80 printer are not the same as for the RX80. But with a little knowledge of what printer control codes mean, it becomes easy to make the required modifications to the *printer driver*. The latter is a piece of software that controls the printing operation. Customis-

170

ing this for a particular printer or situation merely means calling it up and typing in the required parameters. With a printer comes an instruction manual, which at first sight, may seem a little formidable. However, it is worth spending a little time reading it; as it will contain much valuable information.

For example, suppose that you want to print on single sheets. The *end of paper* detector may well frustrate you; as long before the full page of text has been printed, the printer stops and beeps loudly and at the same time the *paper out* light flashes. Some word processors will cope with this situation automatically when single sheet printing is selected. They disable the above detector with software commands. However, not all do this. You are then alone with your printer manual! If you use it properly, you will find that there are two solutions to the dilemma; one uses hardware, the other software. Here they are for the Epson RX80:

(a)   at the rear of the printer chassis are two DIP switches: switch 1, which is an 8-way and switch 2, which is 4-way. The tables that give the functions of these DIP switches are found in the printer manual. It requires only a moment to deduce that on switch 1, position 5 controls the *end of paper* detector. It is factory set to the *active* setting. Flipping it to the other position with the tip of a screwdriver makes it inactive, so preventing it from interfering with your single-sheet printing. The disadvantage is that, since it is a bit fiddly making this change, it tends to be done once only and left in the inactive state (I know, I did it!). All that this implies is that, in the future when printing on continuous stationery, it is wise to keep an eye on the paper supply, since there is now no safeguard if it runs out,

171

(b)   the second method comes to light when one looks through the list of printer control codes. It is found that ESC 8 will turn the paper end detector off, while ESC 9 turns it back on again. Thus, it is possible to include these printer control codes in one's program to permit single sheet printing when required.

This is merely one simple example of how familiarity with the printer manual can be of great help.

Daisy-wheel printers

While the dot-matrix printer is fast and versatile, what it lacks is ultimate quality. In the business world the latter consideration is often far more important than being able to draw pictures, print quickly or switch between various printing effects at will. One answer to this is the daisy-wheel printer, so called because of the plastic or metal wheel which carries the alpha-numeric characters at the tips of its *petals*. As each character needs its own spoke, there are likely to be 96 such petals in the wheel. Photo 9.2 shows a typical daisy-wheel.

The daisy-wheel has an indexed reference position, from which it is spun round until the next character to be printed is aligned with an electromagnetic hammer. This hammer then strikes the tip of the spoke against an inked ribbon, giving an impression on the paper underneath. The direction in which the wheel rotates is always such as to give the shortest path to the required character. As each character is fully preformed, by the very nature of the wheel, there is no compromise in the quality of

Photo 9.2   A typical daisy wheel

the printed letter, numeral, etc. But there is also no way
of changing from one type-face to another except by stopping the printer and changing the daisy-wheel. The
exception to this is that underlining, double strike and
other backspace effects can be called up. To use a daisy-
wheel printer from a word processor while retaining the
facility of mixing the type styles means placing commands into the text that stop the printing operation at
the required points and display a message on screen e.g.
change daisy-wheel now!

Other drawbacks of the daisy-wheel printer are that they are incredibly noisy and extremely slow. Why they are slow is not hard to imagine. Whenever a new character is to be struck, the daisy-wheel has to be spun round to the correct angular position. Even when the shortest path is chosen, this takes valuable time. The complete print head assembly also has to be moved along to the next character position. The hammering of the wheel at a rate of, say, 20 characters per second is responsible for most of the noise, the level for which has been quoted at 75 dB, about twice the acoustic power output of a dot-matrix printer.

One clever feature of daisy wheel printers driven by microprocessors is that the impact force can be related to the nature of the character printed. If, for example, the same force were to be used for printing a capital M and a full-stop, either the M would be too light or the full-stop would perforate the paper. The way in which this is done is to convert a digital value representing the required force into a corresponding solenoid current to drive the hammer. A digital-to-analogue converter performs this function.

Provided that you can live with the minus points mentioned above, what these printers produce on the printed page looks very nice. For draft documents and run of the mill listings they are really not very appropriate. They are essentially for correspondence work only.

Laser printers

The physical arrangement of a laser printer appears in Figure 9.4. The design is based on an electrostatic drum

or belt, which has a photo-sensitive coating. When co-
herent monochromatic light from a helium-neon or
semiconductor laser strikes it at any point, that point
becomes positively charged. This very small point becomes the smallest element of the image to be printed.
The idea is similar to that of the *pixel*, which is the smallest element of the image produced on a television or
computer display. By causing the laser beam to scan
across the drum, line by line, as the drum is rotated, and
turning the laser beam on and off, the electrostatic pattern of dots created forms the image of the page to be
printed.

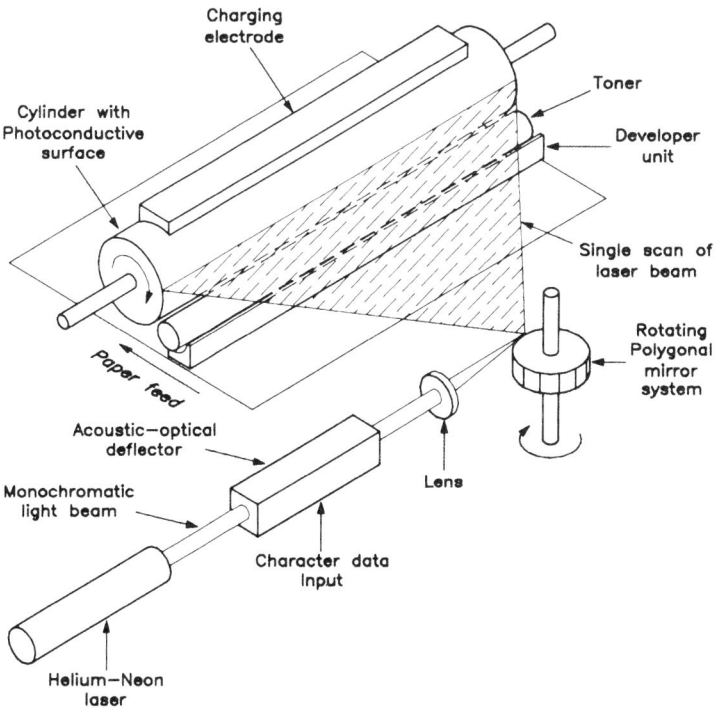

Figure 9.4   Physical arrangement of the laser printer

Computer interfacing

Scanning is performed by using a lens to focus the laser beam on to a rotating polygonal mirror. As the beam hits one face of the mirror, the movement of the mirror causes a single scan of the drum. A regular scan pattern is formed because the drum and mirror rotate in synchronism. The dot-matrix characters are formed by switching the beam on and off under the control of an acoustic-optical deflector in the beam's path. This, in turn, acts according to the character data fed to it. A typical matrix is 18×24 dots; these are overlapped to improve the quality of the printed character.

Because a complete page (say of A4 size) is built up in this way, such printers are often known as *page printers*.

To perform the printing process, the charged drum is next passed over a bath of toner (very fine black powder). The latter is attracted only to those areas on the drum which are positively charged. At this point the sheet of paper is fed into the printing path. This paper sheet has previously been given a very high electrostatic charge, far higher than the charge on the drum itself. As a result, the toner moves from the lower potential of the drum to the higher potential of the paper, thus forming a latent image on the paper. The image becomes permanent when the paper is fed though a pair of heated rollers.

It can be seen from this that the image formed by a laser printer is actually based on a dot pattern rather like that of a dot-matrix printer. The actual method is obviously quite different, giving the laser printer one of its primary characteristics — quietness of operation. However, the dot image of the laser printer is usually better than that of a dot-matrix type because of the greater number of *dots per inch* (dpi). Nonetheless, there is little difference in quality between the top end of the dot-matrix market (24-pin printers) and the bottom end of the laser market. Both may have similar *resolutions*, 300 dpi being typical. This would put the dot size at about 0.085 mm diameter.

The top end of the laser printer market, essential for desk-top publishing (DTP), with machines costing several thousand pounds, produces an image with a resolution four times better, that is 1200 dpi. The corresponding dot size is then approximately 0.02 mm diameter. The laser printer's work rate is fast and may be specified in terms of characters per second (cps) — 400 cps for the Panasonic KX-P4420 — or it may be specified as so many pages per minute, the figure of 8 pages per minute being applicable for the same printer.

The cost of a laser printer does not stop with the initial purchase. Ignoring the paper, which is quite cheap, there are other considerations. The toner refill is quite expensive (£60, say). But there is another factor, that of memory. Many laser printers are supplied with only 512 Kb of memory, insufficient for any *image intensive* work. A single A4 page of graphics requires about 1 Mb of memory! It is wise to make sure that memory can be upgraded and to check the cost of doing so. Other factors to consider are compatibility with the main *page description languages* such as PostScript which interpret the commands given, to generate the dot patterns.

Laser printers come with a range of built-in fonts and more can be down-loaded from the computer. Software

Computer interfacing

support is excellent and, when the prices drop rather more, many more people will doubtless be making use of them.

Inkjet printers

Inkjet printer technology is a form of dot-matrix printing, but the dots are formed by spraying liquid ink through tiny nozzles directly onto the paper. To produce high quality a large number of nozzles are needed, much the same argument as for the other types of dot-matrix printers. One major benefit that comes with inkjet printing is a very substantial drop in the noise level. There are no hammers, naturally! The ink is held in a reservoir from which it is drawn continuously during the printing process. At a price of £15 each, these reservoirs are quite expensive. This means that this type of printer is not really suited to large volume printing. Correspondence and the production of quality *masters* are more likely applications.

One thing that inkjet printers can do pretty well is colour printing. Whereas, in colour dot-matrix printers of a more conventional type, several passes, using a different subtractive colour each time, have to be made, with inkjet printers the required combination of subtractive colours is drawn from the different colour reservoirs simultaneously, through a common print-head. One pass over the paper lays down a solid line of colour. More expensive colour printers have a separate print-head for each colour.

The system for an inkjet printer is shown in Figure 9.5. This shows that the term *spraying* used above, while essentially correct, is actually a simplification. In a little more detail the operation is as follows.

The ink used is actually conductive and is forced through a very fine nozzle, producing a high speed inkjet. The nozzle is vibrated at some ultrasonic frequency, typically 100 kHz, by means of a piezoelectric crystal and crystal driver. The effect of this action is to produce ink drops

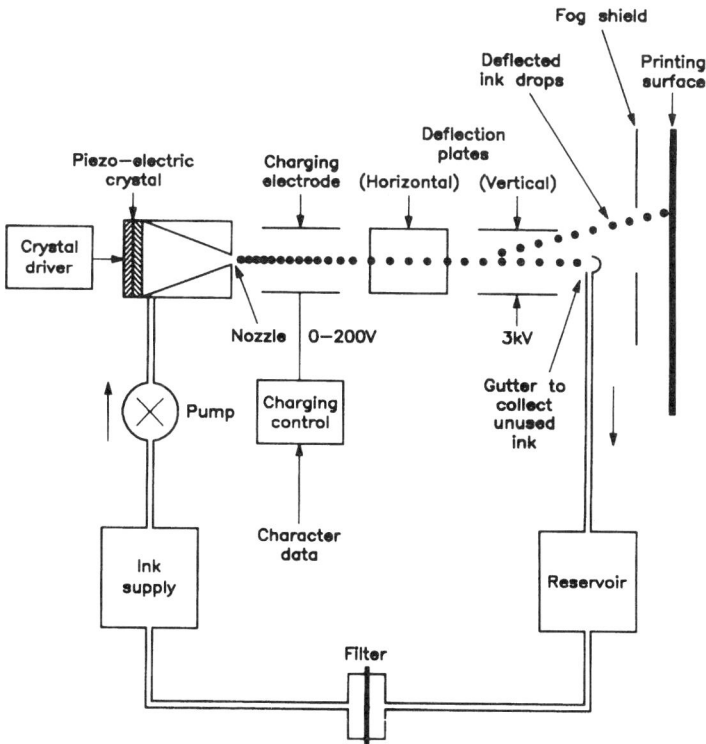

Figure 9.5 The inkjet printer

of constant size, about 0.06 mm in diameter. The ink drops are each then given an electrical charge of specific value. Figure 9.5 shows that this is produced in a charge electrode structure, which is controlled by the data to be printed.

The charged *beam* of ink drops is deflected in a deflection plate assembly, reminiscent of that in a cathode ray tube utilising electrostatic deflection. The horizontal plates may sometimes be omitted if the scan in this plane is by mechanical movement of the mechanism. The degree of deflection is determined by the individual charge on an ink drop. Those that are not charged at all, are collected in a gutter and returned to the ink reservoir.

For one specific case of an inkjet printer, it is stated that it takes 10^3 ink drops to form a single character. With ink drops being generated at the nozzle at the rate of 10^5 per second, this leads to a printing rate of 100 characters per second.

The inkjet printer is a close rival to the laser printer, and like the laser printer, its main advantage is quietness of operation, though price is an important consideration. To approach the quality of laser printers there has to be at least 40 ink nozzles. With the few ink nozzles that some printers have, their performance is no better than many dot-matrix printers.

Notable inkjet printers come from Hewlett-Packard (the HP Deskjet series) and Canon (the BJ Bubblejet series). The former has 50 nozzles giving a resolution of 300 dpi; whileprinting speed is 1.5 pages per minute — the same resolution as for the cheaper laser printers but rather slower. Price is in the £200–300 region.

Thermal printers

Thermal printers were also mentioned in the list of available technologies at the beginning of this chapter. There are some incredibly expensive thermal colour printers, used like the laser printer for page printing. There is another type of thermal printer of the dot-matrix type that requires special paper. Alternatively it may use a thermal ribbon to print onto normal paper. This type of printer has little to offer in the face of competition from the other types of printer and it is doubtful if, as a means of obtaining hard copy from a computer, much real use is made of them nowadays. There is, therefore, little that need be said about them, apart from the above comments.

Printer interfaces

Printers handle their data in serial fashion. This is an obvious statement, since most of them are clearly seen to print each character in turn. Nonetheless, the way in which the data is passed to the printer may be through either a serial or parallel interface.

Essentially there are just two standard interfaces. The serial one is the EIA 232 (or V24) interface covered elsewhere and the parallel one is the Centronics interface. The principles of both serial and parallel data transmission were discussed in reasonable detail in Chapter 5. Briefly however:

Computer interfacing

In the EIA 232 link the TTL levels are converted into two different levels, using negative logic, at the sending end and converted back again at the receiving end. Each character is sent as a separate *data package* framed by start and stop bits. The serial mode uses only a single forward conductor plus a return, unless hardware handshaking (to indicate buffer full) is used and in this case additional conductors will be employed.

The Centronics interface uses eight data lines, plus handshaking lines, plus a large number of ground lines. A standard 36-way Amphenol connector is normally used. Cables connecting computer to printer are usually of necessity quite short, 1–1.5 metres being typical.

In using these interfaces, characters are not actually sent one at a time as they are being printed. In practice, a *stock* of data is held in a small area of RAM termed the *printer buffer*, which is replenished as required through the link.

Index

Computer interfacing

Index

MAPLIN Books

This book is part of a new series developed by Butterworth-Heinemann and Maplin Electronics. These practical guides will offer electronics constructors and students a clear introduction to key topics. The books will also provide projects and design ideas; and plenty of practical information and reference data.

These books are available from all good bookshops, Maplin stores, and direct from Maplin Electronics. In case of difficulty, call Reed Book Services on (0933) 410511.